Lateral Flow Immunoassay

Raphael C. Wong · Harley Y. Tse
Editors

Lateral Flow Immunoassay

Springer

Editors
Raphael C. Wong
Branan Medical Corporation
10015 Muirlands Road
Irvine, CA 92618
USA
raphael@brananmedical.com

Harley Y. Tse
Wayne State University
School of Medicine
Department of Immunology and Microbiology
540 E. Canfield Avenue
Detroit, MI 48201
USA
htse@wayne.edu

ISBN: 978-1-58829-908-6 e-ISBN: 978-1-59745-240-3
DOI 10.1007/978-1-59745-240-3

Library of Congress Control Number: 2008939893

Printed on acid-free paper

springer.com

To our families for their patience and support during the preparation of this book.

This book epitomizes the friendship and collaboration of the co-editors for over forty years.

Preface

Since its initial development in the 1980s, the technology of Lateral Flow Immunoassay has gained wide acceptance. The main reason for its popularity is the simplicity of the test design. The lateral flow immunoassay devices are compact and easily portable. Most of them do not require external reagent for results. Addition of a liquid sample would initiate and complete the test. Results are quick and easy to interpret, usually without the help of an instrument. The technology is also powerful. Multiple analytes can be tested simultaneously with a single device. It can also be easily combined with other technology to provide a comprehensive analysis like simultaneous drug and alcohol determinations by the police force in a roadside testing situation. Manufacturing of the test is relatively easy and inexpensive. Advancement in the detection moieties, improvement in material components, availability of better processing equipment, and increased attention to quality manufacturing all contribute to increase in the reliability, accuracy, and applications of the technology. However, the continuing demand for quantitative result and sensitivity has presented great challenge for assay developer.

In this book, we have compiled the essence of the current state of the art of this technology and addressed what can be expected in the future. After an introduction on the evolution of the technology, some market information is provided. They are followed by discussions on the various materials, biological and chemical components involved in making the test. A chapter on manufacturing equipment and process completes the production section. Later chapters concentrate on the common issues encountered by the users or developers of the technology. Discussions are made on the design of handheld readers, the causes of false results and some way to reduce them, and the regulatory aspects in the development and marketing of lateral flow immunoassay devices.

We hope this book will enhance the understanding on the lateral flow immunoassay technology, enable better products to be made, and provide impetus for further technology advancement.

Irvine, California, USA
Detroit, Michigan, USA

Raphael C. Wong
Harley Y.Tse

Contents

Contributors

Michael C. Brown Strategic Diagnostics, 128 Sandy Drive, Newark, DE 19713, USA, mbrown@sdix.com

Peter Chun, Ph.D., President EASE Laboratories Inc., 338 North Canal Street, Unit 9, South San Francisco, CA 94080, USA, pechun2@yahoo.com

Michael Eberhard ESE GnbH, Ludwigshafenerstr, 29, D-78333, Stockach, Germany, Michael.Eberhard@ese-gmbh.de

Konrad Faulstich, Dr. ESE GmbH, Ludwigshafenerstr, 29, D-78333, Stockach, Germany, Konrad.Faulstich@ese-gmbh.de

Roman Gruler ESE GnbH, Ludwigshafenerstr, 29, D-78333, Stockach, Germany, Roman.Gruler@ese-gmbh.de

Klaus Haberstroh ESE GnbH, Ludwigshafenerstr, 29, D-78333, Stockach, Germany, Klaus.Haberstroh@ese-gmbh.de

Kevin Jones Technical Marketing Manager, Diagnostics, Whatman Inc., 13 Lake Drive, Boonton, NJ 07005, USA, Kevin.jones@whatman.com

Dirk Lentzsch, ESE GnbH Ludwigshafenerstr, 29, D-78333, Stockach, Germany, Dirk.Lentzsch@ese-gmbh.de

Michael A. Mansfield, Ph.D. Diagnostic Business Unit, Millipore Corporation, 80 Ashby Road, MS 75C, Bedford, MA 01730, USA, Michael_mansfield@millipore.com

Brendan O'Farrell, Ph.D. Principal, Diagnostic Consulting Network, 6084 Corte del Cedro, Carlsbad, CA 92011, USA, bofarrell@dcndx.com

Jennifer S. Ponti Director of Sales & Marketing, G&L Precision Die Cutting, Inc., 1766 Junction Avenue, San Jose, CA 95112, USA, jponti@glprecision.com

Shara Rosen, Senior Consultant, StratCom, 1743 Sunnybrooke, Montreal, QC H9B 1R5, Canada, stratcom@total.net

Thomas C. Tisone, Ph.D., President BioDot, Inc., 17781 Sky Park Circle, Irvine, CA 92614, USA, tom.tisone@biodot.com

Harley Y. Tse, Ph.D. Department of Immunology and Microbiology, Wayne State University School of Medicine, 540 E. Canfield Ave., Detroit MI 48201, USA, htse@wayne.edu

Raphael C. Wong, President Branan Medical Corporation, 140, Technology Drive, Suite 400, Irvine, CA 92618, USA, raphael@brananmedical.com

Chapter 1
Evolution in Lateral Flow–Based Immunoassay Systems

Brendan O'Farrell

1.1 Introduction: History of Membrane-Based Point of Care Immunoassays

The development of the rapid, immunochromatographic test strip, also known as lateral flow immunoassay (LFIA), is the result of convergence of several threads that can be traced back to the 1950s. However, the concept of rapid diagnostic tests based on body fluids dated back significantly further. Documented evidence of saliva- and urine-based diagnostics existed several thousand years ago. The ancient Chinese were among the first documented users of saliva-based diagnostics. One widely used practice involved the use of saliva as a rapid determinant of guilt. In the "Rice Test", the inability to generate enough saliva to swallow a handful of rice was considered sufficient evidence for conviction. In this way, a rapid result was generated, but often with a poor prognosis for the subject. One of the earliest written records of a urine-based diagnostic test for pregnancy can be found in ancient Egyptian documents. There, a test was described whereby a potentially pregnant woman could urinate on wheat and barley seeds over the course of several days. The results: "If Barley grows, it means a male child. If wheat grows, it means a female child. If both do not grow, she will not bear at all" [1]. The interest in urine as a rapid diagnostic medium for a variety of ailments continued through the Middle Ages, with the advent of the so-called piss-prophets in Europe, who claimed to be able to differentiate many different conditions from the color of urine. Along with many medical concepts of the time, success typically varied. Despite best efforts through the ages, it was not until the mid-twentieth century that the majority of rapid diagnostic methods gained real predictive value.

The technical basis of the lateral flow immunoassay was derived from the latex agglutination assay, the first of which was developed in 1956 by Plotz and Singer [2]. In the same period, plate-based immunoassays were being developed.

B. O'Farrell (✉)
Diagnostic Consulting Network, Carlsbad, CA 92011
e-mail: bofarrell@dcndx.com

R.C. Wong, H.Y. Tse (eds.), *Lateral Flow Immunoassay*,
DOI 10.1007/978-1-59745-240-3_1, © Humana Press, New York, NY 2009

The first radio-immunoassay (RIA) was invented by Berson and Yalow in the 1950s [3]. The enzyme immunoassay (EIA) was introduced in the 1960s, bringing with it significant advantages, including the replacement of radioisotopes with enzymes, faster reaction times, higher specificities, and longer shelf-life than RIA. The basic principles of the lateral flow technology continued to be refined through the early 1980s and were firmly established during the latter years of that decade, with the filing of several major patents on this technology format by companies such as Becton Dickinson & Co. and Unilever and Carter Wallace [4–6]. Since then, at least another 500 patents have been filed on various aspects of the technology.

The main application driving the early development of the solid phase, rapid-test technology was the human pregnancy test, which represented continual historical interest in urine testing for medical diagnostic purposes. This particular testing application made great strides in the 1970s, as a result of improvements in antibody generation technologies and significant gains in understanding of the biology and detection of human chorionic gonadotropin (hCG), derived largely from the work performed by Vaitukaitis and co-workers [7]. However, to fully develop the lateral flow test platform, a variety of other enabling technologies were also required. These include technologies as diverse as nitrocellulose membrane manufacturing, antibody generation, fluid dispensing and processing equipment, as well as the evolution of a bank of knowledge in development and manufacturing methodologies. All of these elements were required to render a mélange of complex chemicals, biologicals, papers, polymers, people, and processes into a simple and easy-to-use test, which is able to adequately perform to provide prognostic results in a variety of critical applications. Many of these facilitative technologies had evolved throughout the early 1990s, to the point where many are now mature, off-the-shelf technologies. As a result of the early work in all of these areas, the first lateral flow products were introduced to the market in the late 1980s. Since then, the technology, its applications, and the industry have all continued to evolve. As of 2006, over 200 companies worldwide are producing a range of testing formats, with a total value of approximately $2.1 billion dollars (USD) in major market segments (Stratcom, personal communication) (also see Chapter 2). The application of the technology has expanded well beyond clinical diagnostics to areas as diverse as veterinary, agriculture, biowarfare, food, environmental health and safety, industrial testing, as well as newer areas such as molecular diagnostics and theranostics.

The purpose of this chapter is to introduce readers to many of the key elements of the lateral flow immunoassay, to describe the basic process of producing a lateral flow immunoassay, and to understand why it has achieved such broad penetration in so many market areas. This chapter will also discuss the limitations of the current technology and how this technology must evolve to meet ever more demanding market requirements.

1.2 Architecture of a Lateral Flow Immunoassay

Figure 1.1 shows the typical configuration of a lateral flow immunoassay. Traditionally designed assays are composed of a variety of materials, each serving one or more purposes. The parts overlap onto one another and are mounted on a backing card using a pressure-sensitive adhesive. The assay consists of several zones, typically constituted by segments made of different materials. These will be briefly explained here. When a test is run, sample is added to the proximal end of the strip, the sample pad. Here, the sample is treated to make it compatible with the rest of the test. The treated sample migrates through this region to the conjugate pad, where a particulate conjugate has been immobilized. The particle can typically be colloidal gold, or a colored, fluorescent, or paramagnetic monodisperse latex particle (see Chapter 5). This particle has been conjugated to one of the specific biological components of the assay, either antigen or antibody depending on the assay format (see Chapter 4). The sample re-mobilizes the dried conjugate, and the analyte in the sample interacts with the conjugate as both migrate into the next section of the strip, which is the reaction matrix. This reaction matrix is a porous membrane, onto which the other specific biological component of the assay has been immobilized. These are typically proteins, either antibody or antigen, which have been laid down in bands in specific areas of the membrane where they serve to capture the analyte and the conjugate as they migrate by the capture lines. Excess reagents move past the capture lines and are entrapped in the wick or absorbent pad. Results are interpreted on the reaction matrix as the presence or absence of lines of captured conjugate, read either by eye or using a reader. The assay formats can be either direct (sandwich, Fig. 1.2a) or competitive (inhibition, Fig. 1.2b) and should be able to accommodate qualitative, semi-quantitative, and, in limited cases, fully quantitative determinations. Direct assays are typically used when testing for larger analytes with multiple antigenic sites, such as hCG, Dengue antigen, or

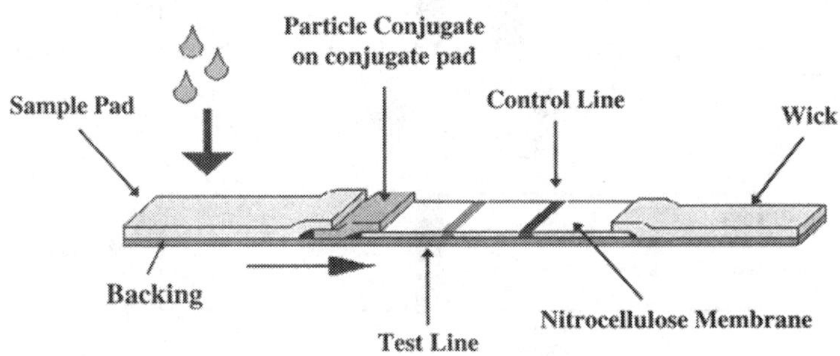

Fig. 1.1 Typical configuration of a lateral flow immunoassay test strip

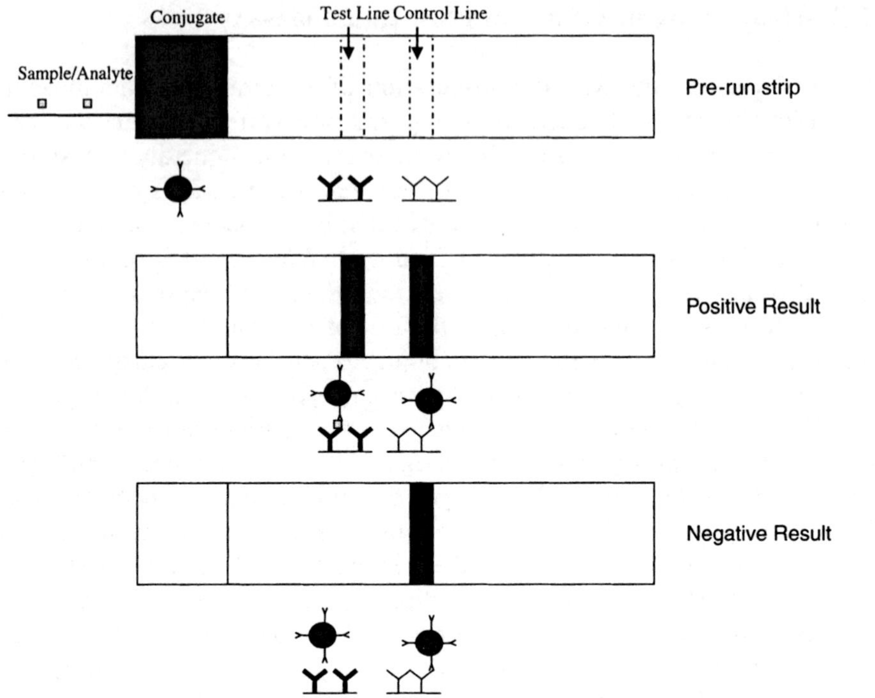

Fig. 1.2a Direct solid-phase immunoassay

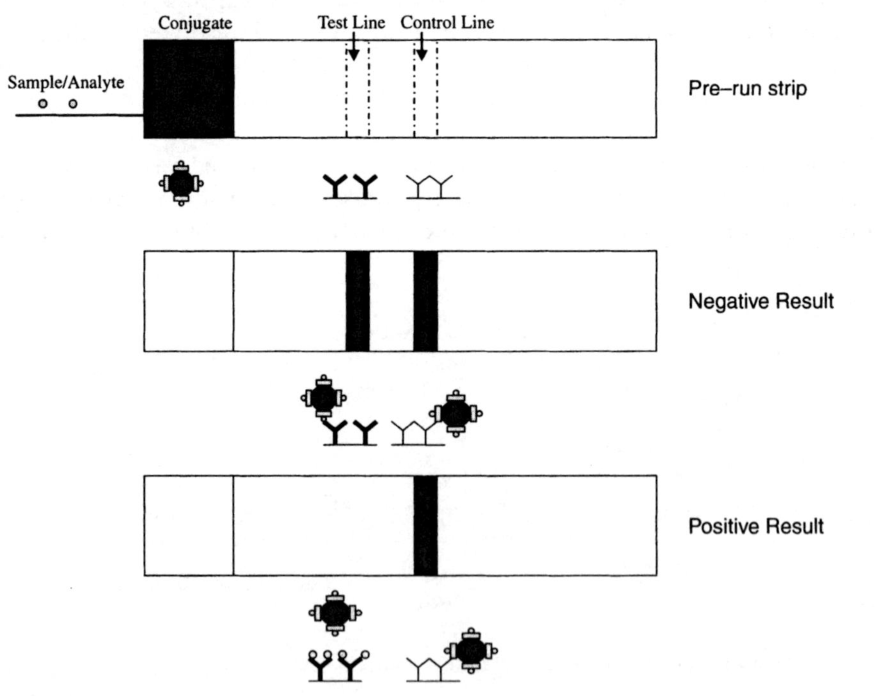

Fig. 1.2b Competitive solid-phase immunoassay

human immunodeficiency virus (HIV). In this case, a positive result is indicated by the presence of a test line. Less than an excess of sample analyte is desired, so that some of the conjugated particles will not be captured at the capture line, and will continue to flow toward the second line of immobilized antibodies, the control line. This control line typically comprises a species-specific anti-immunoglobulin antibody, specific for the antibody in the particulate conjugate. Competitive formats are typically used when testing for small molecules with single antigenic determinants, which cannot bind to two antibodies simultaneously. In this format, a positive result is indicated by the absence of a test line on the reaction matrix. A control line should still form, irrespective of the result on the test line.

1.3 Utility of the Lateral Flow Immunoassay Technology: Advantages and Issues

Lateral flow immunoassays represent a well-established and very appropriate technology when applied to a wide variety of point-of-care (POC) or field use applications. The advantages of the lateral flow immunoassay system (LFIA) are well known:

- Established mature technology
- Relative ease of manufacture – equipment and processes already developed and available
- Easily scalable to high-volume production
- Stable – shelf-lives of 12–24 months often without refrigeration
- Ease of use: minimal operator-dependent steps and interpretation
- Can handle small volumes of multiple sample types
- Can be integrated with onboard electronics, reader systems, and information systems
- Can have high sensitivity, specificity, good stability
- Relatively low cost and short timeline for development and approval
- Market presence and acceptance – minimal education required for users and regulators

Critical among these advantages are the POC nature and a very broad range of applications that can be brought to market extremely quickly and for a relatively small investment. These are advantages that few other putative POC technologies currently in development, including sensor- and array-based technologies, can claim to share. While innovation in microfluidics, biosensor, and multiplexed arrays continues at an increasing rate, those technologies typically require long development cycles, careful market selection, market education, and large investment in technology and infrastructure development in order to make significant impacts in most diagnostic marketplaces.

Traditionally designed lateral flow immunoassays, however, have also suffered from performance limitations, most notably sensitivity and reproducibility. Some of these issues are listed below:

- Unclear patent situation (see Chapter 11)
- Miniaturization of sample volume requirements below microliter level
- Multiplexing: simultaneous analysis of multiple markers difficult
- Integration with onboard electronics and built-in QC functions challenging
- Sensitivity issues in some systems
- Test-to-test reproducibility challenging – limits applications in quantitative systems

These limitations have been exacerbated by the continuing use of traditional manfacturing practices and materials, labels, and visual detection systems. In recent years, market pressures have led to the development of a range of new materials, reagents, detection methods, reader systems, and manufacturing process technologies that together yield the potential to significantly improve the performance of lateral flow immunoassays [5]. Some of these elements will be discussed later in this chapter.

Despite issues of perceived or real performance limitations, lateral flow immunoassays have achieved broad penetration in a variety of markets. Figure 1.3 lists the market segments in which lateral flow immunoassays are already in production or are known to be in development. The manufacturing of lateral flow assays that meet the requirements of some of these market segments, however, may represent new challenges. As applications expand,

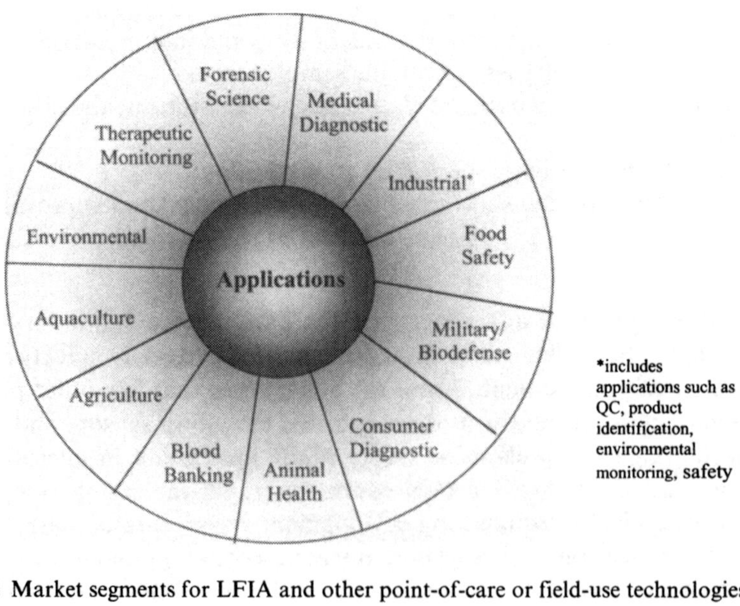

Fig. 1.3 Market segments for LFIA and other point-of-care or field-use technologies

demands on the technology increase, requiring improvements in sensitivity, reproducibility, and manufacturability. Quantification and objective read/record technology, often linked to laboratory information systems (LIS), are being developed. Below is a list of features that are considered desirable for next generation PCOC technologies:

- Rapid and easy to use
- Use small volume of sample, appropriately transferred to the assay without contamination
- Cost-effective to manufacture and use
- Can be manufactured in high volume
- Produce clearly presented and easily interpreted results. Produce high-sensitivity results with low constant of variance (CV)
- Enable quantification
- Integrate to objective read/record technology with built-in connectivity
- Enable multiplexing
- Assays and technology must demonstrate benefit and fulfill a need

In order to meet these demands, there is a growing need for improved materials, assay technology, reader technology, and manufacturing processes. There is also a growing requirement for a more multidisciplinary approach to lateral flow assay development. In the sections below, the elements of the lateral flow technology that present the most challenge to the development of highly reproducible assays will be discussed in the context of the individual components of the assays and the manufacturing methodologies employed in their production.

1.4 Commonly Used Materials and Processes in Lateral Flow Immunoassay Development and Manufacturing

The common components of a lateral flow assay and the standard materials used for each component are considered below. Processing methodologies for each component will also be discussed in relation to how the material drives the manufacturing and development process for typical lateral flow assays.

Figure 1.4 outlines a generalized manufacturing process for traditional lateral flow test strips. The materials and processes typically used for the manufacture of each of these systems and the ways in which the materials are used have remained largely unchanged for much of the history of LFIAs.

1.4.1 Assay Components

1.4.1.1 The Membrane/Analytical Region

Purpose: The purpose of the analytical region in a lateral flow immunoassay is to bind proteins at the test and control areas and to maintain their stability and

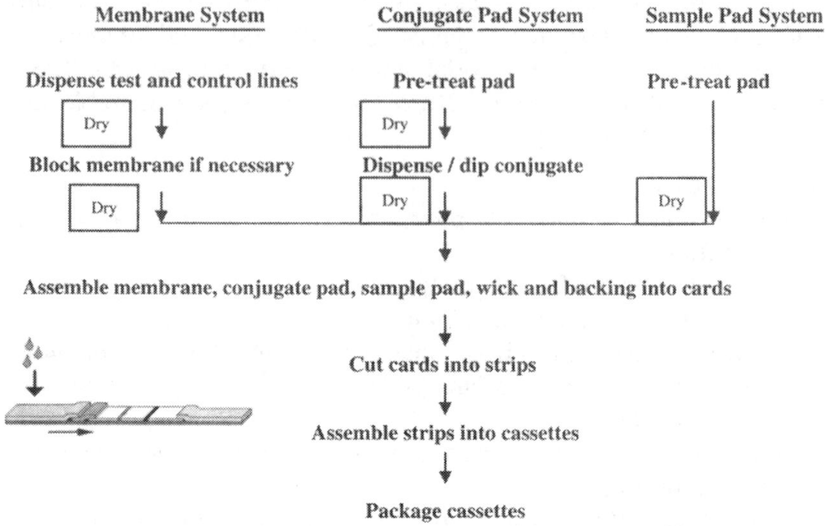

Fig. 1.4 Outline of a typical lateral flow immunoassay manufacturing process [8]

activity over the shelf-life of the product. When the strip is run, it must accept the conjugate and sample from the conjugate pad, flow them consistently to the reaction area, allow the reaction at the test and control lines to happen, and allow excess fluids, label, and reactants to exit without binding.

Material: The material of choice in the vast majority of lateral flow immunoassay systems has historically been nitrocellulose (see Chapter 6). Several attempts have been made to introduce other material types into the market, including nylon and polyvinylidene fluoride (PVDF) membranes. However, those attempts have had limited success, apparently due to factors including cost, limited utility, the need for education regarding new chemistry and processing requirements, and resistance to change due to the large bank of existing experience in the use of nitrocellulose [8].

Nitrocellulose, while extremely functional, is not an ideal matrix for an analytical membrane in LFIAs. It does have certain characteristics that make it useful, and it remains the only material that has been successfully and widely applied in this way to date. These characteristics include relatively low cost, true capillary flow characteristics, high protein-binding capacity, relative ease of handling (with direct cast, or backed membranes), and a variety of available products with varying wicking rates and surfactant contents. However, the material also possesses a variety of characteristics that make it imperfect for this application. These include imperfect reproducibility of performance within and between lots, shelf-life issues, flammability (primarily in unbacked membranes), variable characteristics due to environmental conditions, such as relative humidity and being subject to breakage (if unbacked), compression, and scoring during processing. As a result of these issues with the material, developers and manufacturers spend a considerable amount of time and effort in optimizing chemistries that overcome some of the inherent material issues and in developing manufacturing processes that guarantee adequate performance over the entire shelf-life of the product. Careful control of the key processes of dispensing, dipping, and drying, and attention to chemical and biological treatment of the membrane in order to prevent the introduction of additional variation into the finished product are critical to success.

Flow Characteristics: In order to function as the reaction matrix in a lateral flow immunoassay system, the materials must be hydrophilic and have consistent flow characteristics. Nitrocellulose as a base material is hydrophobic, and is made hydrophilic by the addition of rewetting agents during the membrane production process. These rewetting agents are surfactants, and the type, amount used, and addition methods of surfactant differ from manufacturer to manufacturer and also from brand to brand within a manufacturer. These factors can affect the performance of the assay initially and over time. Not every protein is compatible with every surfactant. This is one reason for screening multiple membrane types during development. The flow characteristics of nitrocellulose membrane change over time, primarily due to desiccation of the membranes upon storage. Nitrocellulose membranes can be envisaged as a sponge, with the pores of the sponge being held open by water. If that water is removed, the pores collapse, disrupting the ability of the membrane to wick fluids through it. This results in changes and inconsistencies in flow rates over time. As speed directly affects assay sensitivity, extended run times can produce false positive results. This is a major contribution to the variability in lateral flow immunoassays.

Critical to the proper performance of a lateral flow immunoassay system is the requirement that it binds reactants only at the desired locations, namely the test and control lines. The protein-binding capacity of a membrane, its interactions with proteins, and the kinetics of the protein-binding process are the parameters that determine the appropriateness of a given set of proteins for the membrane and the sensitivity of the resulting diagnostic tests. Proteins bind to nitrocellulose through a combination of electrostatic, hydrogen, and hydrophobic forces. One of the key elements to the production of sensitive and reproducible assays is the consistent immobilization of immunologically active proteins to test and control lines. It is known that a majority of the proteins lose much of their immunological activity after binding passively to the membrane surface, due to their inability to bind covalently or directionally to nitrocellulose. The commonly accepted model for binding of protein to nitrocellulose is that proteins are initially attracted to the membrane surface by electrostatic attraction. Long-term bonding is then accomplished by a combination of hydrophobic and hydrogen bonds. Many factors affect the binding process, and these must be considered when developing assays and processing nitrocellulose membranes. Some of these factors are listed below:

(i) Reagent choices [9]

- Non-specific proteins: bulking proteins [e.g., bovine serum albumin (BSA), casein] compete for binding sites
- Materials that interfere with hydrogen bonding: Formamide and urea interfere with hydrogen bonding
- Materials that interfere with hydrophobic interactions: Tween and Triton interfere with hydrophobic bonding
- Polymers such as polyvinyl acetate (PVA), polyvinylpyrrolidone (PVP), and poly-ethylene glycol (PEG) interfere with protein binding by a combination of these effects

(ii) Environment

- Humidity should be optimized for binding (25–50% relative humidity at room temperature)

(iii) Processing Methods

- Dispensing methods: Contact tip versus non-contact will have effects on how protein binds or spreads through the membrane
- Drying methods: Forced air oven at elevated temperature versus ambient drying conditions. Drying time and methods can affect the rearrangement and activity of proteins on the membrane

Stability: The membrane must not destabilize bound proteins at test and control lines for entire shelf-life or change its flow characteristics in that period.

Membrane Processing: Nitrocellulose must undergo several processes before integration into the final device. These include deposition of test and control line proteins using quantitative dispensers, drying using forced air ovens at elevated temperature, and immersion processes for blocking. To lay down the test and control line proteins, the membrane is striped with proteins using either contact or non-contact dispensing systems, and is blocked thereafter to control and stabilize the flow-rates and hydration characteristics and to prevent non-specific binding. The dispensing method used for the test and control lines must be as quantitative as possible and should not vary with material hydration or absorption characteristics. Non-contact dispensing methods provide the best solution for quantitatively dispensing proteins onto nitrocellulose. The purpose of blocking a nitrocellulose membrane is to prevent the binding of proteins and labeled conjugate to the membrane at areas other than the test and control lines. Blocking also serves other functions, including maintenance of hydration of membranes, modification of wicking rates, and stabilization of test and control line proteins. Blocking is typically performed by immersion of the membranes in a solution containing proteins, surfactants, and polymers, and is a relatively uncontrolled process. The blocking method must be carefully controlled to produce optimal performance in the final product over its entire shelf-life. Drying is subsequently performed by a combination of blotting to remove surface fluids and with forced air at elevated temperatures to dry. Again, this drying process must be carefully optimized to minimize variation in the final product.

Availability and Choice: The correct combination of membrane types and specific proteins is an important factor for the success of a functional test. Different nitrocellulose membranes can vary considerably in terms of performance characteristics when used with different proteins. Thus, a variety of suppliers and brands of nitrocellulose membranes are available. Performance of the membrane is typically defined by factors such as the polymer type used in the membrane, the pore size, the surfactant type, quantity, and the method of surfactant application. Pore sizes of the membrane used in lateral flow immunoassays range from a nominal 8 to 15 microns, although pore size is a non-exact descriptor in the case of nitrocellulose membranes. The polymeric structure does not actually create pores, but rather a tortuous sponge-like pathway for fluid and particle movement. "Wicking rate" or "capillary rise time" is a more appropriate measure of membrane flow characteristics than pore size. Capillary rise time is defined as the length of time required for a fluid front to traverse a 40 mm width of membrane and is a manufacturer-defined specification for nitrocellulose membranes. The choice of wicking rate is important to the kinetics and speed of development of the assay and will have critical effects on assay performance and sensitivity.

1.4.1.2 The Conjugate Pad

Purpose and General Characteristics: The role of the conjugate pad in a lateral flow immunoassay is to accept the conjugate, hold it stable over its entire shelf-life, and release it efficiently and reproducibly when the assay is run. In practice, variations in conjugate deposition, drying, and release from the membrane constitute major contributions to the coefficient of variation (CV) in assay performance. Assay sensitivity can also be adversely affected by poor conjugate mixing and release from the conjugate pad. Depending on the system, some may favor fast release while others favor slow release of the conjugate. However, the release must always be consistent. Because of the nature of the materials used, it is often necessary to pre-treat conjugate pads to ensure optimal release and stability. Pretreatment is performed by immersion of the pad in aqueous solutions of proteins, surfactants, and polymers, followed by drying. This process, similar to membrane dipping and drying described above, can be performed either in manual batch mode or in continuous inline mode, the latter giving the best opportunity for homogeneous processing of entire batches of materials.

The addition of conjugates to the treated pad is a critical step for the final performance of the test. Two methods are typically used. The first is immersion of the treated conjugate pad into the conjugate suspension. The second is dispensing with quantitative non-contact dispensers such as the BioDot AirJet Quanti 3000 (see Chapter 8, Fig. 8.4). With respect to the conjugate system, the choice of labels and conjugation methods are important. The most commonly used labels include colloidal gold and monodisperse latex, tagged with either a visual or a fluorescent dye (see Chapter 5). The labels can be covalently or passively coupled and can be read quantitatively. Covalent coupling can be crucial to the ability to perform quantitative assays due to the inherently more stable bonds formed between the ligand and the particle as opposed to passive adsorption methods.

Material: The materials of choice are glass fibers, polyesters, or rayons.

Flow Characteristics: For best results, the materials must be hydrophilic and allow rapid flow rates. Most materials used in lateral flow immunoassay systems are very hydrophobic in nature, and must be treated to make them hydrophilic. This is done during the manufacturing of the assay rather than by the material manufacturer, although there are exceptions to that. Most notably are the Accuflow series of pretreated conjugate pads produced by Whatman (Kent, UK). This treatment involves the immersion of the pads in a solution of proteins, polymers, and surfactants, followed by drying at high temperatures as described earlier.

Release Characteristics: The conjugate pad must release the conjugates efficiently and reproducibly over the shelf-life of the product. Typically, some variation in release may occur due to the nature of binding of the particle conjugate to the fibers of the material. It is important during assay optimization to generate stabilization chemistries that minimize this effect and create the most efficient release of particles possible.

Stability: The conjugate pad must not destabilize the conjugate over the entire shelf-life (up to 2 years). Typically, some destabilization does occur, due to the binders present in the majority of these materials. Assay optimization therefore involves the testing of multiple materials for compatibility with the protein–particle conjugate being used.

Manufacturers and Manufacturing Issues: Most commonly used products are from Whatman (Kent, UK), Ahlstrom (Helsinki, Finland), Pall Gelman (East Hills, NY, USA), and Millipore (Bedford, MA). The conjugate pad system is responsible notably for the majority of variations in lateral flow immunoassays. Variation in the material can lead to inconsistent uptake of the pretreatment liquids and conjugates, destabilization of the conjugates, poor release of the conjugate, and binding of conjugates to the hydrophobic fibers. Great care must be taken during manufacturing in optimizing the conjugates, the pad pretreatment process, and the conjugate deposition process to minimize these effects.

1.4.1.3 The Sample Pad

Purpose: One of the major advantages of the lateral flow concept is that these assays can be run in a single step with many different sample types in a variety of application areas. Sample types can be as diverse as whole blood from a post-partum mother, a sputum sample from a potential TB sufferer, or a sample of ground beef from a bulk container. Much of the burden of making those samples compatible with the rest of the assay system falls on the sample application pad. The role of the sample pad is to accept the sample, treat it such that it is compatible with the assay, and release the analyte with high efficiency. Sample treatments include the filtering out of particulates or red blood cells, changing the pH of the sample, actively binding sample components that can interfere with the assay, and disrupting matrix components, such as mucins, in order to release the analyte to the assay. The material chosen to fulfill any or all of these functions can have a great effect on assay performance due to the inhomogeneity of many available materials and the type of binders they contain. The method of pad pretreatment is typically via immersion and drying as with the conjugate pad and, if such treatment is required, the method must be carefully designed to avoid introducing sources of variation, including buffer concentration gradients and edge effects upon drying.

Material: The materials used for the sample pad depend on the requirements of the application. Examples of such materials are cellulose, glass fiber, rayon, and other filtration media.

Capacity: The sample pad material must be treated with assay buffer and other components and dried prior to use. It must also be able to accept all of the sample volume applied to it in a controlled way, thereby helping to channel fluids into the assay materials rather than allowing flooding or surface flow.

Strength: The sample pad material should be strong enough to be handled in manufacturing. An important consideration is tensile strength while wet. If this material is to enter high-volume production, it must endure the tension, without breaking, from a reel-to-reel production system while being immersed in a tank of fluid. The immersion of the pad occurs as part of the pad pretreatment, where the pads are impregnated with an assay buffer containing pH buffer, surfactants, blocking reagents (if required), additives, and other reagents to increase sensitivity of the assay. In some cases, the sample pad and the conjugate pad can be the same unit, although this is not common. Typically, the conjugate and the assay buffer are not compatible. However, it is not unusual to see in some assays the same material being used for the sample pad and the conjugate pad, although the pads are treated individually and subsequently assembled.

1.4.1.4 The Wick

Purpose: The wick is the engine of the strip. It is designed to pull all of the fluid added to the strip into this region and to hold it for the duration of the assay. It should not release this fluid back into the assay or false positives can occur.

Material: The material is typically a high-density cellulose. The choice of wicking material is generally dictated by absorptive capacity, cost, and caliper. Tensile strength and availability in rollstock should also be considerations.

Numerous suppliers of these materials are available, with much of the supply coming from Millipore, Whatman, Ahlstrom, and Pall Gelman.

1.4.1.5 Backing Materials

Purpose: All components of the lateral flow assay are laminated to the backing material to provide rigidity and easy handling of the strip. The backing material is coated with a pressure-sensitive adhesive to hold the various components in place.

Material: The backing materials are typically polystyrene or other plastic materials coated with a medium to high tack adhesive.

Considerations: Incorporation of a backing to a lateral flow immunoassay strip is a necessity in order to laminate multiple materials into one unit performing multiple functions. In traditional, non-reader-based lateral flow immunoassays, the lamination process allows relatively large built-in tolerances in component overlaps and final line placement in a cassette. Variations in overlaps can result in variation in run quality of a strip, although in many less demanding applications this variation can be acceptable. In highly demanding applications, however, variations in run time and fluid front conformation can be fatal to the performance of the assay. This is particularly so in reader-based systems, where the evenness of line development across the entire width of the strip, the speed of running to completion, and the position of the developed line in the assembled cassette can all be absolutely critical to the success of the test. This places high demand in precision in the lamination, cutting, and cassette assembly processes. Automation is key to the success of these processes. The use of inline lamination equipment with camera systems, material edge sensing, as well as sensing in cutting and assembly, are all feasible approaches. Consideration should also be given to the tendency of the adhesive to flow into other components of the test strip, notably unbacked membranes, conjugate pads, and sample pads. This can cause disruption of flow patterns and the creation of hydrophobic patches, as well as destabilization of proteins. For this reason, medium, rather than high, tack adhesives are typically used. Care must be taken to choose adhesives that have a track record of being compatible with proteins during storage of the test strips.

Suppliers: G&L, Adhesives Research (Glen Rock, PA). The majority of the market uses G&L backing cards with GL187 pressure-sensitive adhesive.

1.4.1.6 Labels for Detection

The most commonly used particulate detector reagents in lateral flow systems are colloidal gold and monodisperse latex. Latex particles coupled with a variety of detector reagents, such as colored dyes, fluorescent dyes, and magnetic or paramagnetic components, are available commercially. Detailed discussions of these reagents can be found elsewhere in this book (see Chapter 5).

The choice of the particulate label and the detector reagent used in a particular lateral flow system is driven by a variety of factors:

i) Is covalent attachment of the protein to the particle required?

If this is a requirement, then activated latex particles only can be used, as binding of proteins to colloidal gold is typically achieved via passive absorption.

ii) Is the assay intended to be quantitative or qualitative? Is a reader required?

The development of truly quantitative lateral flow immunoassays requires a great deal of effort in choosing the basic materials and technologies to be used. Some of these considerations will be discussed later in this chapter. In addition, the integration of readers into these assay systems will also be discussed. If the application requires a reader technology, then the choices of label expand to include colloidal gold, colored or fluorescent latex, and paramagnetic latex particles. The choice of particle and reading technology is driven by the consideration of the cost of the reader technology, the cost and availability of licensing of the readers and the labeling technology, the cost of assay development and reader integration, as well as the performance of the combined reading and labeling technology.

iii) What are the levels of sensitivity required?

In visually read assays, it is often possible to generate more sensitivity using colloidal gold rather than colored latex particles, due to the smaller size of the gold particles and, as a result, higher packing density can be achieved on a test line. Gold particles are typically in the range of 20–40 nm in size whereas colored latex particles are about 100–300 μm. Gold also has a higher color intensity than colored latex particles, which allows for better discrimination of low positives in an assay. On the other hand, latexes can be produced in multiple colors and can utilize darker colors such as dark blue dyes to provide greater contrast against the white background of a lateral flow membrane. In reader-based assays, it is often possible to generate even higher sensitivity by using fluorescent particles or paramagnetic particles [10, 11].

iv) Is the assay intended to be multiplexed? Is discrimination between multiple lines of different colors required?

If these parameters are required, the choices are reduced to colored latex particles, which can be produced in multiple colors.

v) What are the cost considerations in the manufacturing processes?

There are several major suppliers of colloidal gold in the marketplace, such as British Biocell International (BBI, Cardiff) and Diagnostic Consulting Network (DCN, Carlsbad, CA). In general, purchasing colloidal gold is more economical than producing gold internally. It is possible to integrate the process of colloid gold production into the lateral flow immunoassay manufacturing process, and numerous companies have chosen to do this. However, when consistency of the product and the quality of the assay results are considered important, the overall goal to reduce any sources of variation in the assay often leads companies to contract with dedicated suppliers of high-quality components. There are also multiple sources of the monodisperse latex, most of which are considered to be of high quality, including Bangs Laboratories and Merck/Estapor. The manufacturing of monodisperse latexes is generally not integrated into a lateral flow immunoassay production process. As a result, companies relying on outside supply also have cost considerations.

1.4.2 Processing Methodologies

A comparison of traditional and improved manufacturing technologies will be presented in a separate chapter of this book (see Chapter 8). Only an outline will be given here. The basic processing steps as shown in Fig. 1.4 involve dispensing of reagents, immersion of materials into bulk solutions of reagents, drying of components, lamination of materials, cutting into strips, and packaging. Process designs can be broadly categorized into two methods. In batch processing, card lengths of materials are processed individually, assembled into cards, and cut into strips. In in-line or reel-to-reel processing, all components are maintained in roll format until they have been treated and laminated, and only then are they cut into either individual strip or card lengths for final packaging. It is intuitive and widely supported by industry data [12] that inline processing significantly reduces manufacturing variation in lateral flow immunoassays.

In the final analysis, strip-to-strip variation is among the top concerns expressed by the clinical users of lateral flow technologies (Stratcom, unpublished market research). The reduction of these variations, coupled with the ability to integrate the assays to data collection and reporting systems, will be the key to bringing this technology to a broader range of applications.

1.5 Improving the Utility and Performance of Lateral Flow Immunoassays: Trends in Devleopment of New Technologies

This section considers some of the reasons why much of the potential of rapid membrane testing technology remains untapped, and where this technology may be headed.

The application of lateral flow immunoassays covers multiple market segments with widely diverse performance requirements, market forces, and commercialization strategies. As a result, it is not possible to generate a single definition for the attributes of "next generation" point-of-need assays. However, there are recognized trends and desired design attributes for the development of new tests. Some are met by current designs and approaches to market, while others will require radically different approaches. Manufacturers looking to develop successful point-of-need products face three challenges. First, they must select and implement features and benefits that cost-effectively match the needs of end users. Second, they must develop the core technologies necessary to create a functional product. And finally, they must ensure that they have selected appropriate market applications for the core technology [13]. The market forces that drive the acceptance of a product in a particular market segment will be discussed elsewhere in this book (see Chapter 8). The remainder of this chapter will concentrate on the discussions of core technology developments considered likely to provide improvement for lateral flow type devices.

1.5.1 Necessary Improvements in Performance Based on Core Assay Technology

Since 1995, over 500 patents have been issued in areas directly related to lateral flow immunoassay technology covering everything from sample pretreatment to improvements in matrices, cassette designs, label types, processing methods, result interpretation, and many others. It is the view of users and developers alike that the technological keys to achieving better market penetration of a point-of-need assay are improvements in test-to-test variation and in sensitivity. Current levels test-to-test variations have prevented the performance of quantitative assays that approach those of larger or more complex clinical analyzers. Improvements in sensitivity would also allow assay systems to be applied in areas where larger clinical immunoassay systems, and methodologies such as PCR, are considered the gold standards. Other general technology features that are considered critical include the integration of reader technologies and data capture systems.

The sensitivity requirement differs significantly depending on the assay and on the application. One consideration is the definition of sensitivity. There are two main definitions applicable to point-of-need assays. The first is "response per unit ligand". This is the slope of the dose–response curve and is primarily applicable in quantitative assays. The second is "the lowest level of non-zero ligand reliability". This is measured as the lowest detectable dose (LDD), which is applicable in qualitative assays. Which definition to use depends on the characteristics of the assay. The LDD is the most commonly applied definition in lateral flow immunoassays, where the aim is to distinguish affected from unaffected members of a population. The unit response, however, becomes important when we are considering truly quantitative assays. Both measures are dependent on the slope of the dose–response curve. A steep slope on the curve provides a better unit response and lower LDD than a shallow curve. This in effect defines the discriminatory ability of the assay.

1.5.1.1 Application of Alternative Materials

It is generally accepted that novel materials are needed to improve the functional performance of lateral flow immunoassays. Materials that can perform multiple functions are desirable [11]. The application of new materials opens the door to entirely new concepts of device design and potential levels of performance for LFIA. Some of the preferable characteristics of an improved matrix for point-of-need assays are listed below:

- Highly regular surface, yielding cosmetically high-quality lines
- Three-dimensional matrix with consistent pore size, thickness, and protein-binding capacity
- True capillary flow with a variety of wicking rates
- Thin as reasonably possible
- Good fluid flow characteristics over its entire shelf-life, independent of treatment
- Low CVs for capillary rise time over its entire shelf-life, independent of treatment
- Minimal metal contaminants
- Low background fluorescence
- Non-interfering
- Stable on storage
- Non-flammable
- Low cost
- Can be activated for covalent linkage
- Multiple functionality: can act as conjugate application area, sample application area, reaction surface, separation medium, and wick all in one

Several new materials and approaches that take these basic principles into consideration are under development. Some examples include Fusion 5 from Whatman and the 4CastChip from Amic, Sweden.

i) Fusion 5

One attempt at creating a matrix that meets at least some of the above specifications is the Fusion 5 matrix (Whatman) (see Chapter 7). This is a large-pore, single-layer matrix. It is hydrophilic and non-protein binding in nature. This material fulfills all of the required functionalities of the components in the traditional lateral flow device, namely sample pad, conjugate pad, membrane, and wick. Because the material is non-protein binding, a traditional method of applying the test and control lines onto Fusion 5 is not possible. A strategy of laying down "boulders in the stream" is therefore applied [9]. Large-diameter beads (approximately 2 μm) are used to conjugate to specific proteins and are dispensed onto appropriate locations on the material, where the large-size beads become immobilized and form the test and control line areas. When the test samples and protein conjugates flow past the "boulders", binding and signal formation occur at those locations. The open pore nature of this system means that assays can be run extremely fast, which have both positive and negative effects. Speed or reaction time can be important to LFIAs, but can also be related inversely to sensitivity in many instances. With Fusion 5, this is largely overcome due to the lower inherent background and the increased surface area for ligand binding provided by the beads.

The use of Fusion 5 can reduce complexity in the manufacturing process. Multiple dipping, drying, and lamination steps are no longer applicable. The major challenge to the broad application of Fusion 5 as a material is the fact that it comes in only one flow rate, which is very fast. There are no choices in terms of the other material characteristics.

ii) CastChip

A second approach to generate novel assay and material design comes from Amic (Uppsala, Sweden). Amic has developed an assay substrate that consists of a highly ordered array of micropillars on a plastic slide. These micropillars are hydrophilized by dextran, and act to drive capillary flow of sample and reagents in the flowpath. The pillars also provide a biocompatible surface for the attachment of capture ligands at test and control lines. The material is highly regular compared to a standard nitrocellulose material. As with the Fusion 5 approach, the Amic device contains the capacity for multiple functionalities, as the pillar-defined flowpath can act as sample application area, reaction surface, and wick.

Binding of proteins to the surface of the substrate is by covalent attachment via amine linkages to the aldehyde groups on the chip surface. Protein is dispensed onto the surface and then allowed to react in a humid environment for a short time for the linkage to occur. This substrate can generate sensitive assays using fluorescent labels and can be linked to a reader system.

In terms of manufacturing, this device, like the Fusion 5, removes several steps from the process. However, it also introduces the need for discontinuous dispensing of proteins onto discrete substrates, and this can only be achieved in a non-contact manner. From a processing standpoint, the requirement for individual handling of chips is a drawback relative to the ability to handle materials such as nitrocellulose in an inline fashion. Dispensing methods must be highly regular, reproducible, and carefully controlled to ensure that line widths are consistent.

The 4CastChip represents a shift in thinking in the context of LFIA substrates. It is effectively a two-dimensional substrate without discreet pores. However, the device does exhibit true capillary flow; has an extremely regular, hydrophilic surface; and generates cosmetically acceptable lines at sensitivities in systems tested to date that are comparable to existing clinical and POC systems. The material itself is highly stable and maintains proteins in stable conditions for extended periods. As such, it meets many of the criteria previously listed. One challenge to the broad applicability of this system is the cost, which remains high relative to nitrocellulose. Nevertheless, the chip manufacturing technology has the capacity to become very inexpensive with volume, as it is based on CD molding technology.

Fusion 5 and 4CastChip are examples of novel approaches to substrate design, which may or may not ultimately achieve broad market penetration or importance. What is critical, however, is that they represent possible alternatives specifically for application in point-of-need assays. This is a trend that continues to grow, particularly as companies from outside the traditional diagnostic material–supply industries become more interested in this market.

1.5.2 Evolution in Design

It has been argued that the future of the lateral flow technology is inextricably linked to developments in biochip technology (Stratcom, unpublished market research). To some degree, this is true in the broadest sense. In the long term, it is possible that true displacement technology, based on the principles of microfluidics and single-molecule detection, will displace technologies such as the lateral flow device. However, the very characteristics that have defined the broad success of lateral flow – ease of manufacture, low cost and short timeline to development, ease of use and interpretation, broad applicability across sample types and markets, and user and regulatory acceptance – are not inherent in current approaches to biochip and multiplexed and miniaturized assays. The vast majority of these formats still require the development of facilitative technologies for their production and have huge obstacles in terms of proving performance and generating market acceptance. The lessons of i-Stat attempting to introduce new decentralized testing technologies in clinical markets should be carefully studied by those moving into this area. It required years of effort, hundreds of millions of dollars, severe regulatory hurdling, and the development of a whole system of clinical support and distribution to thoroughly break the i-Stat technology into the market. By comparison, up-grading the performance of lateral flow based systems is likely to be much simpler, involving improvements of materials, manufacturing processes, and the application of newer design concepts that are likely to provide faster access to market, with lower costs and burdens.

1.5.2.1 Use of Alternative Conjugate Integration Strategies to Improve Variation Issues

It has been demonstrated that a large component of inter-assay variation derives from the act of impregnating solid substrates with particulate conjugates and drying the conjugates in place before subsequent release using the fluid phase of the assay [14]. As a result, alternative methods for the introduction of conjugate to the assay system can have the result of decreasing assay–assay variation. This has been demonstrated by, for example, Response Biomedical's Rapid Analyte Measurement Platform (RAMP), where the conjugate is premixed with the sample and thereafter delivered directly to the strip along with the sample. This is one element of RAMP's assay design and result interpretation strategy, particularly the use of the ratio of control to test line intensity ("the RAMP Ratio"), which leads to lower than average strip–strip variation.

1.5.2.2 Integration of Microfluidics to Lateral Flow Systems

Another mechanical element leading to variation in lateral flow immunoassay results is the accurate delivery of sample to the strip and the subsequent movement of the sample and conjugates through the device to the reaction matrix. Several attempts have been made to utilize fluidic elements to overcome the issue of accurate sample delivery. One example is the HemaStrip design, originally developed at Saliva Diagnostic Systems in Vancouver, WA, which utilized a capillary collector molded to the end of a plastic tube in which the strip rests.

1.5.2.3 Sample-Handling Considerations for Assay Sensitivity

For many applications, clinical and otherwise, traditional lateral flow formats and labels are capable of providing significant sensitivity. However, there is a growing trend in certain clinical applications, such as detection of cardiac markers, to attempt to achieve assay sensitivity that would previously have been considered beyond the ability of a point-of-need test to deliver. Other areas that require the same level of extreme sensitivity include biowarfare applications, such as anthrax detection, and food microbiology, where single-organism detection is the goal. Standard approaches of labeling and detection in lateral flow are unlikely to reach the required sensitivities for these applications. Instead, the industry must look to alternate labeling approaches, coupled with the use of reader systems. However, it must be noted that with diagnostic devices, the label and the reader are only part of the equation. The appealing feature of lateral flow and other point-of-need assay systems is that they provide a complete "sample-to-answer" solution in a single step. Simply developing a high-sensitivity reader and labeling system is not enough. It is critical to consider the system as a whole, including the sample, the sampling method, the sample pretreatment methodology, and the concentration of analytes. Analyte concentration can be a confounding factor when it is either too high or too low for detection. There are also considerations of the affinity of the antibodies (see Chapter 4) and how their activities can be affected via the conjugation methods used during labeling procedures. These factors must also be considered when entering into a design program to create highly sensitive and highly reproducible assays. These concepts are illustrated by the following examples:

i) High concentration of analyte: The high dose hook effect is a well-known phenomenon in assay development. For an immunoassay to give accurate results, there must be an excess of antibodies, both capture and label, relative to the analyte being detected. Only under conditions of antibody excess does the dose–response curve show a positive slope and provide accurate quantitation. As the concentration of analyte begins to exceed the amount of antibody, the dose–response curve will plateau. Further increase of analyte will cause the slope to become negative. Care must be taken in assay development to validate all potential sample types by dilutional linearity analysis to establish if they are on the positive slope region of the curve. The choice of label, conjugation method, and reading method will all have effects on the ability of the assay to handle large ranges of analyte concentrations. Sample pretreatment may in some cases be necessary to reduce analyte concentration prior to the assay to prevent hook effects.

ii) Low concentration of analyte: Analytical methods are traditionally divided into several steps: sampling, preliminary operations, measurement, calculation, and evaluation of results [15]. Sampling refers to the generation of a representative sample of an inhomogeneous object [16]. This inhomogeneity presents a challenge to the success of the analytical method. A perfect example of the issues brought about by the quest for the ultimate sensitivity – single-molecule detection – was discussed in an interview with Graham Lidgard of Nanogen published in IVD Technologies in May 2006 [17]. Dr. Lidgard took the example of an assay that was developed by Tomas Hirschfeld of the Los Alamos National Laboratory (Los Alamos, NM) [18] in which the researchers coupled a polymer of fluorescein to an antibody and directed a laser through a microscope into the sample. When the molecule passed under the laser, instant photobleaching occurred. The problem was that if there was only a single molecule in solution, it could take three months for this molecule to pass in front of the laser. Therefore, to generate results in a reasonable length of time, several hundred thousand molecules were required to be present in the solution. In a related paper [19], Chen et al. concluded that "the fluctuation in the number of molecules taken for chemical analysis is a fundamental and irreducible source of uncertainty.... [and that] the inhomogeneity [of the sampled solution] presents a fundamental limit to analysis". This same uncertainty applies to the sampling of biological or other matrices for immunodiagnostic testing. As it applies to highly sensitive point-of-need assays, the moral of this tale is that it is not the absolute sensitivity of the system that is the most critical factor. Rather, it is the ability to acquire as representative a sample as possible, and ultimately, it is the concentration of the analyte detectable in the primary sample that is critical. Sampling and pretreatment methods are therefore critical in determining the availability of many analytes in an assay. If one takes 100 ml of a homogenized food sample and concentrates it into 100 μl for analysis, the assay system will have significantly more sensitivity than one that takes 100 μl of that primary sample without concentration and tests it. All the sensitivity benefits are there without ever changing a label or a detection method.

iii) High incidence of cross contaminants in samples (see Chapter 10): Certain assays are well known for their high false positivity rates, including *Mycobacterium tuberculosis* (TB) and *Chlamydia trachomatis*. The Chlamydia assay, where high numbers of organisms are often present, does not require high sensitivity. Rather, it requires antibodies of high specificity. The TB assay has both sensitivity and specificity challenges. The sample pretreatment method, the antibody selection method, and the labeling and reading methods must be designed around these issues.

1.5.3 Reader Systems in Lateral Flow Assays

Reader technologies employed in lateral flow applications (see Chapter 9) are based on one of three labeling and detection technologies: Detection of colloidal gold or colored monodisperse latex particles using charged coupled device (CCD) cameras, detection of fluorescent monodisperse latex particles based on LED excitation via confocal or other optical sensors, and detection of paramagnetic monodisperse latexes using Magnetic Assay Reader (MAR™) technology. It is beyond the scope of this chapter to discuss these technologies in detail, but numerous references are available [10, 11, 23].

The implementation of reader systems in marketed lateral flow systems has been accomplished only sparingly to date. The Biosite Triage and Response Biomedical RAMP systems are among the only examples currently on the market for quantitative applications, and the Biosite device does not utilize a standard lateral flow design. Other readers are used for qualitative assays, including drugs-of-abuse assay systems, such as Cozart's DDS™ or Rapiscan™ products, American BioMedica Corporation's RapidReader™ for their RapidScreen™, and other drugs-of-abuse assay formats. Non-clinical applications also use visual readers, such as Neogen's Reveal Accuscan, which accepts their lateral flow tests for Listeria, *Escherichia coli* O157:H7 and *Salmonella*, and the company's thin test strips for GMOs, mycotoxins, food allergens, and ruminant material. Recently, "in-cassette" readers have been utilized in several products. However, quantification of results is not the intent of these systems either. These devices serve as an analog to digital conversion of a yes/no result, with the purpose of removing user interpretation errors (e.g. the "Clearblue Easy"™ from Inverness Medical).

Several factors can account for the limited application of reader devices in the lateral flow immunoassays:

i) Most applications of LFIAs have utilized visual labels which do not require readers. Among the many positive attributes of LFIA applications is simplicity. It does not require the cost and complexity of a reader to generate results.

ii) LFIA's have been traditionally applied in areas where results recording and data capture are not required.
iii) Quantitative LFIA systems have proven extremely difficult to produce due to variability issues with the assays rather than the readers.
iv) For reader-integrated device development, access to or development of an appropriate reader system is a major issue. Reader systems are not readily available to developers of lateral flow immunoassays. They are beyond the ability of most lateral flow manufacturers and developers to produce. Limited options for original equipment manufacturers (OEM) of reader systems have traditionally existed. Furthermore, integration of reader systems adds complexity to product design and development over that of traditional LFIA products. Many lateral flow developers do not have the expertise or experience to develop an integrated system and take it through regulatory channels. The cost of such development, even with many of the OEM producers of readers, is inhibitory to most small- and mid-sized lateral flow developers.

Given the fact that readers do not define the ultimate sensitivity of the assay, and that the chemistry and biology ultimately provide the result, where should a developer put the most effort? Clearly, the chemistry and biology of lateral flow immunoassay must be made to work in association with the reader and the sample-handling methodologies. This is the area that provides the most fertile opportunity for assay development companies, the majority of which are unlikely to become directly involved in reader development due to issues of cost and lack of expertise. Several companies provide readers on an OEM basis to the market. The Leach Technology Group manufactures visual reader systems for inclusion in OEM platform products. Members of the Leach Group are Hauser Inc., Westlake Village, CA; UMM Electronics Inc., Indianapolis, IN; and LRE Technology Partner GmbH, Munich and Nordlingen, Germany. The major entract to the field of OEM reader supply in recent times has been ESE GmbH (Stockach, Germany), with a range of portable benchtop and handheld fluorescent and visual readers that can be integrated with products quickly and cheaply (also see Chapter 9). An example of a fluorescent and visual lateral flow immunoassay reader from ESE is shown in Fig. 1.5. These readers are expected to greatly reduce the complexity and cost of quantitative assay development in the near- to mid-term and should encourage entrants to this field.

It is expected that market requirements will continue to drive this move toward integrated reader technologies for lateral flow immunoassays. Market drivers include the following aspects:

- The effort to push many diagnostic tests out of the central lab into the home and point of need continues. Data capture, appropriate result interpretation with minimal user error, and the opportunity for data mining are all important features of that trend. These requirements will feed the need for reader systems at the point of need, even for qualitative assays. In clinical

Fig. 1.5 ESEQuant portable benchtop lateral flow strip reader by ESE GmbH. Fluorescent or visual labels can be read

applications, integration of the point-of-need assay with the hospital's information system will be critical to the acceptance and use of the assay by physicians.

- The ability to connect assay results at the point of need with other elements of the healthcare system will make result interpretation, data monitoring and storage, and transduction of data into action feasible at centralized sites with feedback capability.
- Increased sensitivity will require non-visually interpreted results.
- Quantification requires a reader.

1.5.3.1 Consideration of Reader Design for Lateral Flow Applications

Cost, Maintenance, and Calibration: A reader should not be over-engineered. In terms of cost, the consensus among manufacturers in the clinical area is that they will need to give readers away or provide them at low end user cost. So the cost of the reader becomes critical. Abbott established the "razor and blades marketing strategy" in the mid-1980s, with the launch of the first truly automated immunoanalyzer – the TDx. Thereafter, this became the strategy followed by most companies that market tests on small readers (Source: Stratcom). In terms of maintenance requirements, users of readers in point-of-need applications will not tolerate a requirement for regular maintenance or calibration of a reader. As a result, the readers should be simple, robust, and subject to internal or on-strip calibration procedures. This is one strong argument for the application of single-use devices such as those seen in the Clearblue Easy hCG test (Inverness Medical).

Communication/Connectivity: The reader should be able to collect, store, and communicate data to other sources. The technologies by which this can be accomplished, including Bluetooth and USB, are now standard across multiple industries and relatively simple to implement for reader manufacturers. The requirement for this feature cannot be over-emphasized in any market segment to which the readers will be applied. The human clinical market is a good example. Pressure for having device connectivity comes not only from multiple sources, including physicians themselves, but also from the US government and external industry sources. The US Health and Human Services Office has called for a National Health Network to enable healthcare providers to securely share the myriad of data, records, and images on patients. The aim is to provide seamless healthcare services across systems, doctor's practices, clinics, labs, and hospitals. Compliance to various regulations such as the Health Insurance Portability and Accountability Act (HIPAA) notwithstanding, this effort is estimated to be a $200 billion dollar market [20]. Major consortia are being formed to address the challenges of this industry, including the Continua Alliance, with such notable members as Cisco Systems, IBM, Oracle, CSC, Microsoft, Hewlett-Packard, and Intel. The presence and interest of companies like these, through organizations like the Continua Alliance, guarantee that connectivity will play an increasingly important role in diagnostics. The integration of reader systems with onboard connectivity into assays will become one of the major trends in point-of-need diagnostics in the twenty-first century.

Throughput: The reader methodology, coupled with how the assay is run, defines the potential throughput of the assay system. Due to the limitations of lateral flow design, and the fact that these are not end-point assays, control of the time at which assays are read can be critical to the generation of appropriate test results in quantitative assays. As a result, certain reader designs used in marketed assays require that the assays be run on the reader, with timing of result reading done by the reader. In essence, that severely limits the throughput of the system. Depending on the application, high throughput may be more or less of an issue; but in applications such as clinical laboratories where tests may be batched before being run, this will be a critical issue. The inclusion of multiple read-heads will be one element that can overcome this concern. However, the ultimate answer lies in the test design and result interpretation algorithms.

Sensitivity: The reader must be calibrated to cover the entire dynamic range of the assay. In certain instances, such as fluorescence, this may require desensitization of the reader relative to potential performance, as available signal may swamp the detector.

Footprint and Portability: Small size and portability may be a positive or negative attribute depending on the market area being addressed. Assuming the unit has onboard connectivity, there is no technological requirement for lateral flow readers to be large, bench-top units. The footprint and size and feel of the reader will be defined by the requirements of different markets.

1.5.4 Examples of Novel Label and Reader Systems for Next-Generation Applications

As discussed previously, the primary labels used in lateral flow immunoassay are colloidal gold and mondisperse latex, labeled with colored, fluorescent, or magnetic tags. The major suppliers of colloidal gold are Diagnostic Consulting Network (DCN) and British Biocell International (BBI). Dyed latexes and paramagnetic particles are available from a variety of sources including Bangs Laboratories, Dynal, Merck/Estapor, and Magsphere. There are also a variety of other labels and enhancement methods that are in development for use in point-of-need assays. One example is nanoparticles. Magnetic, latex, metal, and semiconductor particles on the nanometer scale have unique optical, electronic, and structural properties that can be used in a variety of clinical applications. However, the promised advances in performance based on nanotechnology have yet to materialize in the point-of-need industry. Numerous labeling technologies are in development, but it may yet be many years before some of these technologies produce improvements to existing labeling technologies in real-world applications. As yet, there are few examples of any of these labels in marketed point-of-need assays, particularly in lateral flow format. Examples of newer label technologies can be identified:

i) Up-converting Phosphor Technology: STC Technologies, Inc. (Bethlehem, PA), now Orasure Technologies, developed the up-converting phosphor technology (UPT). UPT is based on lanthanide-containing submicrometer-sized ceramic particles that can absorb infrared light and emit visible light. They have the advantage that biological matrices do not up-convert. Additionally, the particles do not photobleach and are inert to common assay interferents such as hemoglobin. Orasure Technologies, Inc. developed a reader system based on these particles called the Uplink, and demonstrated its utility in example systems including one for *E. coli* O157: H7 [21]. The label has yet to achieve broad penetration in the market.
ii) Quantum Dots: Quantum Dot Corporation (Palo Alto, CA), now owned by Invitrogen, originally commercialized the quantum dot (Qdot) nanocrystal technology developed at California's Lawrence Berkeley National Laboratory, the Massachusetts Institute of Technology, and the University of Melbourne (Source: Stratcom). Qdots are 10–20 nm-sized crystals containing a few hundred to a few thousand atoms of cadmium mixed with selenium or tellurium, which have been coated with an additional semiconductor shell (zinc sulfide) to improve the optical properties of the material. The size of the nanocrystal determines the color of the excitation achieved when it is excited with a long-wavelength UV lamp. As a result, multiplexed assays can be created when conjugates of different sized particles are used as labels in the assay system and are excited using a single wavelength [22].

iii) SERS Tags: Oxonica (Mountain View, CA) produces the Nanoplex biomarker detection system, which utilizes silica-coated, surface-enhanced Raman Scattering (SERS)-active metal nanoparticles as labels for bioconjugates. As with the QDot , the Nanoplex particles boast the ability to produce multiple excitation wavelengths from a single excitation frequency, allowing their use in multiplexed assays [23].

1.6 New and Growing Applications of Lateral Flow Point-of-Need Assays

1.6.1 Point-of-Need Nucleic Acid Tests

Driven by the need to detect nucleic acid products and the opportunity to increase test sensitivity in point-of-need environments, there is a strong market desire to couple nucleic acid detection technology to the speed and convenience of the lateral flow immunoassay. The key to this type of technology lies in the ability to produce highly sensitive results in a very short period of time.

The detection of nucleic acid product in point-of-need environments has been difficult due to a variety of factors. Technical difficulties associated with the most commonly used amplification technology, polymerase chain reaction (PCR), revolve around the challenges of sample preparation, power demands, and control of assay specificity and reproducibility. Complex solutions using automated sample-handling systems coupled with microfluidics and biosensor detection methods are in development, but are not conducive to point-of-need applications. As a result, simpler, alternative strategies are being sought to improve both the amplification and the detection methods. Recombinase Polymerase Amplification (RPA) developed by ASM Scientific (Cambridge, UK) represents a significant step forward in the reduction of the complexity of nucleic acid amplification methods to the point where they can be applied in point-of-need environments [26]. However, lateral flow technologies have not yet been widely applied in this area. A variety of modifications to the standard approach can allow for the application of this technology in Nucleic Acid Lateral Flow (NALF) and be the final step in bringing this key application directly to the market.

A number of strategies are available for the detection of nucleic acids in lateral flow systems [26–29]. The capture of nucleic acids can be performed in an antibody-dependent or antibody-independent way. For example, Piepenburg et al. [27] describe an antibody-dependent system, where an anti-biotin antibody immobilized on the surface of nitrocellulose is used to capture biotin and carboxyfluorescein (FAM) bearing oligonucleotides in RPA amplicons. Binding is subsequently detected using an anti-FAM-colloidal gold conjugate. An antibody-independent alternative utilizes streptavidin as the binding agent. Immobilization of oligonucleotide probes directly onto membranes is also possible using oligonucleotides linked to carrier proteins. Still another possible configuration is to use the "boulders in the stream" approach, where the oligonucleotides are immobilized onto the surface of large latex beads. The beads are subsequently dispensed onto the reaction matrix, forming the usual test and control lines. This methodology serves to vastly increase the available surface area for binding of analyte, thereby improving the sensitivity of the assay. In terms of detection on a NALF strip, the same labels can be employed as on a standard lateral flow immunoassay, including colloidal gold, colored, and fluorescent monodisperse latex particles. Similarly, the basic elements of the manufacturing and equipment can be readily applied to nucleic acid lateral flow. However, simply applying the basics as currently practiced will also mean applying many of the sources of variation and limitation of the lateral flow technologies. Rational improvements to the basic technologies will be key to unlocking the full potential of this assay format in nucleic acid applications.

1.6.2 Proteomics, Therapeutic Monitoring/Theranomics

The link between proteomics and diagnostics is a logical and potentially critical one. A clear need exists for new diagnostic targets for such pathologies as ovarian cancer, bladder cancer, pancreatic cancer, and Alzheimer's disease. It is hoped that proteomics will be able to identify such targets. Even though current commercial activities in proteomics are focused on developing analytical technologies, there have been increasing efforts to develop clinical applications as well [30]. An example of the successful use of proteomics coupled with diagnostic tools was illustrated in a paper by Drydna et al. [31]. Using 2D electrophoresis and mass spectrometry, these researchers identified a protein that could differentiate between rheumatoid arthritis and osteoarthritis. An ELISA was subsequently developed for clinical use based on this marker. An important aspect of this work is that it also provided a potential means for monitoring therapy, thereby linking the theranostics concept. By isolating chronic disease markers that are subjected to the influence of therapeutic drugs, the entire loop of discovery, diagnostics, therapeutics, and monitoring is utilized. In this scheme, self-monitoring or doctors' office monitoring of therapy using rapid assays makes sense on many levels. This is especially so if point-of-need assays can be made to be quantitative and reproducible. The outcome will likely require the utilization of reader and data recording strategies discussed earlier in this chapter. Another example of appropriate application of this strategy is in cardiac diseases. There is currently an active search for coagulation and lipid metabolism proteins that will help improve cardiac risk assessment. In 2005, the world market for these enzyme and protein markers was $20 million. Most assays are developed for specific drug discovery projects and have not been commercialized on a large scale. Between 2005 and 2010, these tests will gain importance and it is predicted that by 2010 this market segment should grow to $40 million (Source: Stratcom).

The determination of susceptibility to disease conditions or the probabilities of therapeutic successes is a possible application of rapid tests. In instances where pathologies may have resulted from changes in several proteins, either generative or consequential, diagnostic strategies will require multiplexing and subsequent analysis using algorithms in a process known as profiling [32]. For these applications, lateral flow type assays may not be appropriate as multiplexed systems such as microarrays are more likely to be utilized. However, for simpler systems that rely on the detection of only a single or relatively few proteins, rapid assays based on lateral flow principles may be applied.

1.6.3 Infectious and Chronic Disease

In considering the worldwide market applicability of diagnostics, a socio-economic division is often applied. The world is somewhat arbitrarily divided into First, Second, and Third Worlds, and population, disease states, and diagnostic applications tend to be broadly considered along those lines. This classification ignores the heterogeneity of population and conditions present within each of those arbitrary boundaries. Cardiac and other chronic diseases in the expanding middle classes of emerging economies are growing, as are the incidences of previously geographically limited infectious diseases (e.g., malaria, dengue), emerging diseases (e.g., H5N1 Influenza), and heretofore well-controlled diseases (e.g., TB in First World Countries) in developed countries. Twenty well-known diseases (e.g., cholera, malaria, TB) have reemerged or spread geographically since 1973, often in more virulent and drug-resistant forms. At least 30 previously unknown disease agents have been identified since 1973, including HIV, Ebola, hepatitis C, and SARS (Source: Stratcom). As a result of this globalization of disease states, one of the most critical issues for diagnostic companies is the development of appropriate distribution and sales strategies in worldwide markets, as well as the navigation of local regulatory and legal conditions. In chronic diseases, there remains significant growth particularly in the areas of inflammation, cardiac markers, and cancer, with a myriad of new labels in development in the search for improved diagnostic and prognostic indicators.

1.6.4 Non-human Applications

The application of lateral flow immunoassays beyond the clinical human diagnostics market is also continuing to grow. Examples of areas of applications are listed below:

i) Animal Health: e.g., equine pregnancy, bovine pregnancy and fertility, and companion animal infectious disease testing.
ii) Agriculture: e.g., genetically modified organisms (GMO) detection, crop quality testing.
iii) Biowarfare: anthrax detection
iv) Environmental and Health and Safety: e.g., contaminating enzymes in manufacturing plants; Legionnaire's disease in air conditioning and water systems.
v) Food Microbiology: e.g., *E. coli* O157, *Salmonella*, *Listeria*, and other food spoilage organisms.

1.7 Conclusions

Lateral flow immunoassay technology is evolving rapidly. Novel approaches driven by market needs are leading to improvements in performance and utility to a vast array of new application areas. With the integration of new reading, labeling, sample-handling, and device designs comes a requirement for a new approach to system development and manufacturing. The development of highly sensitive and reproducible/quantitative next-generation point-of-need diagnostic assays requires a different, more multidisciplinary approach than has been the case with standard lateral flow immunoassays. Input is required from a range of disciplines, including materials science, chemistry, biology, optics, software and hardware engineering, as well as process design, equipment design, and project management. For this reason, a more collaborative approach is required, and companies such as Diagnostic Consulting Network are established with the purpose of fulfilling the many needs of developers in this complex area.

All of the novel approaches discussed in this chapter, as well as the improvements to standard approaches discussed in another article in this series (see Chapter 8), have ramifications for the developers and manufacturers of LFIA and will require changes in the way we think of LFIA. The acid test for any of these technologies will be market acceptance, which in turn will be driven by

- Relevance to the application and the end user
- Manufacturability of the product
- Cost of the product
- Availability for licensing of the individual technologies, as well as the cost of licensing
- Clear license for use, free of other patent infringement

Careful application of lateral flow technologies in well-chosen market areas, coupled with robust, simple reading technologies, novel materials, the correct labels, modified device designs, and appropriate manufacturing strategies, will drive the acceptance of this technology in a vast array of application areas.

Through continuous improvement and evolution in design and performance, lateral flow principles can be applied in ways that have the potential to create entirely new paradigms in high-sensitivity point-of-need testing and will be applied in the market for a long time to come.

References

1. http://history.nih.gov/exhibits/thinblueline/timeline.html#1970
2. Plotz, C.M. and Singer, J.M. (1956) The latex fixation test. Application to the serologic diagnosis of rheumatoid arthritis. Am. J. Med. 21(6):888–892.
3. Berson, S.A. and Yalow, R.S. (1959) Quantitative aspects of the reaction between insulin and insulin binding antibody. J. Clin. Invest. 38:1996–2016.
4. Campbell, R.L., Wagner, D.B. and O'Connel, J.P. (1987) Solid-phase assay with visual readout, US Pat. 4,703,017.
5. Rosenstein, R.W. and Bloomster, T.G. (1989) Solid-phase assay employing capillary flow, US Pat. 4,855,240.
6. May, K., Prior, M.E. and Richards, I. (1997) Capillary immunoassay and device therefore comprising mobilizable particulate labeled reagents, US Pat. 5,622,871.
7. Vaitukaitis, J.L., Braunstein, G.D. and Ross, G.T. (1972) A radioimmunoassay which specifically measures human chorionic gonadotropin in the presence of human luteinizing hormone. J. Obstet. Gynecol. 15:751–758.
8. O'Farrell, B. and Bauer, J. (2006) Developing highly sensitive, more reproducible lateral flow assays. Part 1: New approaches to old problems. IVD Technology, June issue, p. 41.
9. Jones, K.D. (2006) Membrane-based tests. The Latex Course Proceedings, October.
10. Laborde, R. and O'Farrell, B. (2002) Paramagnetic particle detection in lateral flow assays. IVD Technology, April issue, p. 36.
11. O'Farrell, B. and Bauer, J. (2006) Developing highly sensitive, more reproducible lateral flow assays. Part 2: New challenges with new approaches. IVD Technology, July issue, p. 67.
12. Tisone, T. (2000) In-line manufacturing for lateral flow diagnostics. IVD Technology, May issue, p. 43.
13. MacFarlane, I. and Davis, F. (2002) Building blocks for the point of care boom. IVD Technology, January issue, p. 27.
14. O'Farrell, B. (2006) Developing approaches to the development and manufacture of highly sensitive, reproducible lateral flow assays. Proceedings of the Oak Ridge National Conference.
15. Harris, W.E. and Kratchovil, B. (1981) An introduction to chemical analysis. Saunders College Publishing, New York, pp. 4–6.
16. Kratchovil, B. and Taylor, J.K. (1981) Sampling for chemical analysis. Anal. Chem. 53(8):924A–938A.
17. Lidgard, G. and Park, R. (2006) Simplifying detection technologies. IVD Technologies, May issue, pp. 28–33.
18. Hirschfeld, T. (1976) Limits of analysis. Anal. Chem. 48(1):16A–31A.
19. Chen, D.Y. and Dovichi, N.J. (1996) Single-molecule detection in capillary electrophoresis: molecular shot noice as a fundamental limit to chemical analysis. Anal. Chem. 68:690–696.
20. www.continuaalliance.com
21. Niedbala, R.S., Feindt, H., Kardos, K., Vail, T., Burton, J., Bielska, B., Li, D., Milunic, D., Bourdelle, P. and Vallejo, R. (2001) Detection of analytes by immunoassay using up-converting phosphor technology. Anal. Biochem. 293(1):22–30.
22. http: //probes.invitrogen.com/products/qdot/overview.html
23. http://www.oxonica.com/healthcare/healthcare_biodiagnostics.php

24. Bonenberger, J. and Doumanas, M. (2006) Overcoming sensitivity limitations of lateral flow immunoassays with a novel labeling technique. IVD Technology, May, pp. 41–46.
25. Davies, C. (1994) Immunoassay Design. In "The Immunoassay Handbook", D. Wild. Ed., Stockton Press Publisher, New York, Pp. 15–48.
26. Seal, J., Braven, H. and Wallace, P. (2006) Point of care nucleic acid tests. IVD Technology, November, p. 41.
27. Piepenburg, O., Williams, C.H., Stemple, D.L. and Armes, N.A. (2006) DNA detection using recombination proteins. PLOS Biol. 4(7):001–007.
28. Dineva, M.A., Candotti, D., Fletcher-Brown, F., Allain, J.-P. and Lee, H. (2005) Simultaneous visual detection of multiple viral amplicons by dipstick assay. J. Clin. Microbiol. 43(8):4015–4021.
29. O'Farrell, B. (2007) Sensitive, specific and rapid Nucleic Acid Detection at the Point of Need using simple, membrane-based assays. BioWorld Europe, March 2007, 36–39.
30. Lundblad, R.L. and Wagner, P.M. (2005) The potential of proteomics in developing diagnostics. IVD Technology, March issue, pp. 20–22.
31. Drydna, S., Ringel, B., Kekow, M., Kuhne, C., Drynda, A., Glocker, M.O., Thiesen, H.-J. and Kekow, J. (2004) Proteome analysis reveals disease associated marker proteing to differentiate RA patients from other inflammatory joint diseases with the potential to monitor anti-TNF alpha therapy. Pathol. Res. Pract. 200:165–171.
32. Gillespie, J.W., Gannot, G., Tangrea, M.A., Ahram, M., Best, C.J.M., Bichsel, V.E., Petricoin, E.F., Emmert-Buck, M.R. and Chuaqui, R.F. (2004) Molecular profiling of cancer. Toxicol. Pathol. 32(Supp 1):67–71.

Chapter 2
Market Trends in Lateral Flow Immunoassays

Shara Rosen

2.1 Introduction

Immunoassays have been used in hospitals, laboratory medicine, and research since the mid-1960s. In industry, immunoassays are used to detect contaminants in food and water, and in quality control to monitor specific molecules used during product processing. Immunoassays are performed in central laboratories, using a variety of instrument-based technologies or on-site via rapid test techniques.

Rapid immunoassays commonly come in two configurations. One is lateral flow, a one-step technique where the sample is placed on a test device and the results are read in 5–30 min. Lateral flow devices can test for a single analyte or multiple analytes. The second is a flow-through system, which requires a number of steps – placing the sample on the device, a washing step, and then adding an analyte–colloidal gold conjugate that makes the test result visible to the eye. The results can then be read after a few minutes, but the whole process can take as long as 20 min. One sample is tested per cassette. This format is less popular than lateral flow because of the necessity for multiple assay steps and the greater skill that is required for operating these devices.

Lateral flow immunoassay tests, also known as immunochromatographic strip tests, have been a popular platform for rapid immunoassays since their introduction in the mid-1980s. The definition of lateral flow (LF) used here is broad and includes any manual- or instrument-read immunochromatographic strip test/device for a single analyte or multiple analytes that is in a strip format or housed in a cartridge, that uses a paper, nitrocellulose, or plastic support, and that is based on fluid migration or flow technology. This broad definition is used because LF tests have been developed using combinations of all of these features. The remainder of the chapter uses this broad definition to describe the LF format.

S. Rosen (✉)
Senior Consultant, StratCom, 1743 Sunnybrooke, Montreal, QC H9B 1R5, Canada
e-mail: stratcom@total.net

R.C. Wong, H.Y. Tse (eds.), *Lateral Flow Immunoassay*,
DOI 10.1007/978-1-59745-240-3_2, © Humana Press, New York, NY 2009

The LF format is so versatile that manufacturers of rapid immunoassay tests have developed LF tests for almost any situation where a rapid test is required. In hospitals, clinics, physician offices, and clinical laboratories, LF-based tests are used for the qualitative and quantitative detection of a wide variety of antigens and antibodies. More recently, they are being developed to measure the products of gene amplification. They can be used to test just about any biological sample, including urine, tears, sweat, saliva, serum, plasma, whole blood, and biopsied tissue and fluids. LF-based tests are also used in veterinary medicine, for quality control, to assure product safety in food production, environmental testing, water safety, and pharmaceutical and biologics manufacturing. In these circumstances, rapid tests are used to screen for pathogens and toxins in raw materials, the manufacturing environment, and the finished product. Water utilities test drinking water for chemical toxins, metals, and pathogens. Rapid test systems are also used to assess progress in environmental remediation of soil and ground water pollutants. Veterinarians use rapid tests to screen commercial livestock and household pets for a number of medical conditions.

Another reason for the continued popularity of LF tests is the low development cost and ease of production. If the target analyte and necessary antibodies are available, assay development costs can be as low as $30,000 or not higher than $100,000. Further, prototype test kits can be available for clinical trials in as little as 4 months. Once developed and tested, test kits can be manufactured for $0.10–$3.00 per test.

The remainder of this chapter describes the world market for rapid immunoassays and LF tests, and discusses the progress that LF tests have made in immunoassay test segments. Further, the chapter addresses the impact that patents and new technologies will have on the future of LF-based tests.

2.2 Market Size and Trend

2.2.1 Lateral Flow Segments

The world market for LF-based tests [1] is estimated at $2,270 million in 2005 and, with a compounded annual growth rate (CAGR) of 10%, it will reach $3,652 million in 2012 (Table 2.1). This estimate includes LF-based tests used in human and veterinary medicine, food and beverage manufacturing, pharmaceutical, medical biologics and personal care product production, environmental remediation, and water utilities. LF tests are also available and in development for biowarfare agents and pathogens such as anthrax, smallpox, avian influenza, and other potential biological weapons. These are not included in this market analysis because these tests are not in routine use at this time and are in fact a market in waiting mode.

Growth in LF testing is derived mainly from the clinical and veterinary sectors where more of the current test menu will be used in more places outside

Table 2.1 World LF test revenues (in $ Millions)

	Sales 2007	% Market	Sales 2012	% Market	% CAGR	Number of Companies
Clinical	2,010	89	3,240	89	10	50–100
Veterinary	205	9	310	8	9	at least 50
Food & Beverage	30	1	60	2	15	approx. 20
Pharma/Biologics	15	0.6	30	1	15	approx. 20
Environment	5	0.2	6	0.2	4	at least 50
Water Utilities	5	0.2	6	0.2	4	approx. 20
	2,270	100	3,650	100	10	

Source: Kalorama Information, company reports

a central laboratory in response to increased interest in public health issues and the prevalence of chronic diseases. Since the early 1990s, clinical LF tests have grown in popularity. Predictions are that patient self-testing will sky-rocket because of rising consumer expectations, technological innovations, and the surge of consumer activism in healthcare. Furthermore, pharmacies, retail outlets, and physician offices are establishing their position for patient wellness screening. The expectation is that, under these conditions, point-of-care (POC) testing – self-tests and professional – will grow at 20–25% per year instead of the 10% predicted here. However, this may be more a case of wishful thinking than reality. This is because current tests and technologies cannot accommodate the needs of the consumer and professional markets. For consumer self-testing, many of the tests are blood-based, which is not a user-friendly sampling type. Thus, manufacturers have begun the search for alternate samples such as urine and saliva. In the professional setting, most POC tests do not meet the quality standards offered by laboratory-based tests. Further, the thought is that new tests and technologies are just too expensive. Faster, more sensitive, more user-friendly, and less expensive tests may produce better market penetration.

Generally speaking, as is shown in Table 2.2, LF tests so far have had limited success in penetrating the market for immunoassays. Worldwide, even clinical LF tests have managed no more than 33% market share compared to their lab-based immunoassay counterparts. The technology issues discussed above are only part of the reason. The main issue is that they compete with established lab-based test strategies that are more sensitive and less expensive when direct material costs are considered.

2.2.2 Lateral Flow Geography

Worldwide, there is a huge demand for decentralized rapid tests. The US market accounts for $1,005 million (44%), the European market approximately 35% ($799 million), and the rest of the world (ROW) the remaining 21% ($466

Table 2.2 LF test market penetration, worldwide, 2007 (in $ Millions)

	Lateral flow	All	% Market
Clinical	2,010	5,790	35
Veterinary	205	520	39
Food & Beverage	30	290	10
Pharma/Biologics	15	210	7
Environment	5	100	5
Water Utilities	5	20	25
	2,270	6,930	33

All industrial = lab-based EIA, ELISA (not hygiene) for microorganisms/toxins
All clinical, veterinary = lab-based immunoassays that also have a rapid segment
Source: Kalorama Information

Table 2.3 World LF test revenues, by Region (in $ Millions)

	Sales 2007	US	EU	ROW
Clinical	2,010	865	705	440
Veterinary	205	105	80	20
Food & Beverage	30	23	6	1
Pharma/Biologics	15	8	4	3
Environment	5	2	2	1
Water Utilities	5	2	2	1
	2,270	1,005	799	466

Source: Kalorama Information and company reports

million) of the worldwide market (Table 2.3). The European Union (EU) is not a single market but is a confederation of countries, which, for the purposes of this discussion, includes the original 13 EU countries: Austria, Belgium, Finland, France, Germany, Greece, Ireland, Italy, Luxembourg, Netherlands, Portugal, Spain, and UK.

As the largest single market for rapid test products, the United States has an enormous impact on how the rapid test industry develops. In healthcare, food safety, and biologicals manufacturing, the pressures of cost and quality management are pushing the need for rapid and decentralized testing. However, even in the United States, POC and LF in particular have so far not lived up to expectation. This aspect of LF technology is discussed in more detail below.

2.2.3 Lateral Flow Tests in Veterinary Medicine

Lateral flow tests are used in veterinary medicine to test commercial livestock (large animals – cows, poultry, sheep, pigs, etc.) and household pets (cats, dogs, etc.) for a variety of medical conditions including: bacterial and viral infections, allergies, fertility issues, and diabetes. This test segment is not to be confused with food safety testing, which is normally performed on animals that have entered the food chain.

Approximately 60% of immunoassays for small and large animals are performed in reference laboratories. In 2007, LF-type tests occupy some 39% of the veterinary rapid immunoassay and ELISA tests performed. In 2007, these tests generated revenues of $205 million worldwide and, with 9% CAGR, will reach $310 million in 2012. However, as large animal veterinarians are expected to contain outbreaks of infectious diseases in livestock, they are beginning to rely on newer LF tests for more rapid test results. In the area of small animal veterinary practices, rapid on-site testing provides significant advantage in this extremely competitive market. In all cases, the animal owner pays for these tests out-of-pocket. Further, testing in the large animal segment is somewhat regulated by ministries of food and agriculture. The small animal segment is totally non-regulated.

The market is highly fragmented with different tests used for specific animal species. Some 60% of the LF veterinarian market is occupied by tests for household pets. These tests include: tests for anemia, cancer, endocrine function; viral pathogens such as canine adenovirus type-1 (CAV-1), calicivirus (CaCV), adenovirus, coronavirus, parvovirus, rabies, rotavirus, West Nile Virus, etc.; and bacteria and parasites such as *E. coli*, cryptosporidia, heartworm, hookworm, leishmania, leptospirosis, Lyme disease, roundworm, and tapeworm.

No one company dominates the market for LF tests in veterinary medicine. The leaders are Idexx Laboratories, Inc. (Westbrook, ME), Neogen, Inc. (Lansing, MI), and Heska Corp. (Loveland, CO). Other significant players include Agen Biomedical (Acacia Ridge, Australia), Biogal-Galed Labs (M.P. Megiddo, Israel), and Biomedica GmbH (Vienna, Austria).

2.2.4 Lateral Flow Tests in the Food and Beverage Industry

In the past 3–5 years, food safety issues and concerns for public health have led to more stringent legislation in food safety requirements. Legislation has produced increased demand for pathogen and toxin tests in just about every segment of the food production industry – processed food, meats, poultry, beverages, and dairy; and by all major food producers worldwide. In 2007, most food safety tests are performed using lab-based traditional microbiology techniques, but the appreciation of rapid on-the-spot tests is on the rise. These tests are used primarily for testing raw materials, interim products in the manufacturing process, and final product for bacteria such as listeria, salmonella, *E. coli*, and others.

Traditional microbiological testing requires finished product to sit in a warehouse pending test results that can take two to four days, whereas rapid technologies provide microbial test results in as little as 24 hours. The financial benefit of rapid microbial testing methods is a necessity in today's competitive environment.

The food and beverage sector (food) is the largest segment within the industrial rapid test market. In 2007, LF-based tests generated $30 million and, with a CAGR of 15%, will reach $60 million in 2012. However, in 2007, LF tests for food and beverage testing account for only 10% of all immunoassays used in this sector (See Table 2.2).

There is a growing demand from food companies for quicker testing so as to release finished goods more quickly and thus reduce inventories. But, despite their potential for saving inventory and warehousing costs, LF tests account for only a small share of the total number of food microbiology tests. Tests used for food safety are highly regulated and most regulatory bodies have not established formal rapid test certification procedures. In Europe, it is only Denmark (the Danish system of validation of microbiological test methods called the DanVal) and France (the Association Francaise de Normalisation – AFNOR), which have established product approvals, and the Association of Analytical Communities (AOAC) approves tests in the United States.

A driver in the demand for rapid and LF tests in food production is the adoption of Hazard Analysis and Critical Control Points (HAACP) regulations that prescribe test procedures throughout the manufacturing process. The combination of increased competition within the European food industry and the launching of the European validation system, MICROVAL, for alternative food testing methods will make it possible for Europe, within the next few years, to catch up with the leading position of the United States in the use of rapid food testing methods. In addition, large food exporting countries such as Mexico and Brazil, as well as growing processed food industries in Asia Pacific, will provide another avenue of growth for rapid tests.

In a move to facilitate LF test use on-site in smaller, low-volume facilities and those that do not already own immunoassay instrumentation, manufacturers market pregnancy test-like immunoassay dipsticks for *Salmonella*, *Listeria*, and *E. coli*. No one company dominates the market for LF food tests. The leaders are Strategic Diagnostics, Inc. (Newark, DE), Neogen, Idexx Labs and Biocontrol Systems (Brownsville, CA). Other companies include Celsis International PLC (Chicago, IL), Medical Wire & Equipment Co. (Wiltshire, United Kingdom), Merck KgaA (Dermstadt, Germany), and M-Tech Diagnostics Ltd. (Cheshire, United Kingdom).

2.2.5 Lateral Flow Tests in the Pharmaceutical Industry

Similar to the food industry, product safety is a major concern in the production of pharmaceuticals, medical biologicals, and personal care products (known as "pharma"). There is no specific regulatory body that oversees the sterility of these pharma products. However, pathogen and toxin content can impact on product quality, which then detracts from a company's brand image and eventually profit earning potential. Thus, pharma product companies have been quick to recognize the contribution that rapid and LF tests can make to their bottom line. Pharma products are manufactured with a long shelf-life, sometimes a year or more. They may be contaminated with slow-growing bacteria, yeasts, and molds that may be of little consequence early in a product's shelf-life, but can destroy product quality once they have grown.

Unfortunately, the pharmaceutical industry generally has been hesitant to introduce innovative systems into the manufacturing sector and so LF tests have managed to capture only 7% of the market for immunoassays in this sector (See Table 2.2). The main reason is that there has been no regulatory imperative to do more testing.

This situation may soon change, due to the Food and Drug Administration's sponsorship of a voluntary regulatory process. In 2004, the FDA issued a guidance document intended to describe a regulatory framework (Process Analytical Technology, PAT) that will encourage the voluntary development and implementation of new quality assurance procedures. The guidance [2] was prepared by the Office of Pharmaceutical Science in the Center for Drug Evaluation and Research (CDER), under the direction of FDA's Process Analytical Technology (PAT) Steering Committee with membership from CDER, Center for Veterinary Medicine (CVM), and Office of Regulatory Affairs (ORA).

Similar to HAACP, PAT is intended to be a system for designing, analyzing, and controlling manufacturing through the timely measurement of critical quality and performance attributes throughout the manufacturing process – from raw materials to final product.

No one company dominates the market for LF pharma tests. The leaders are Strategic Diagnostics, Neogen, Idexx Labs, and Biocontrol Systems. Other companies include: Celsis International PLC, Medical Wire & Equipment Co., Merck KGaA, and M-Tech Diagnostics Ltd.

2.2.6 Lateral Flow Tests for Environmental Remediation and Water Testing

Environmental remediation and drinking water quality are highly regulated by government agencies in all developed countries and some of the more prosperous developing countries. Agencies such as the EPA, FDA, and the Food Safety and Inspection Service of the US Department of Agriculture test environmental samples and, together with the Association of Official Analytical Chemists International (AOAC), establish official methods for use in testing for environmental contaminants in certain market segments. The AOAC also manages procedures and guidelines for validating new methods, worldwide.

Throughout the OECD, countries have enacted safe water legislation such as the EU Directives and the US Safe Drinking Water Act. Regulations differ from country to country, but overall they are expected to minimize water contamination by industries by regulating the point sources that discharge pollutants into lakes and rivers.

Because of the significant impact that contamination can have on public health, most testing is therefore performed in certified laboratories by approved methods. LF and other rapid tests are performed primarily for spot checks in between required test periods. They are also used by households to verify the cleanliness of well water. As such, this is a small market that together account for not more than $10 million, worldwide. In this area, LF and rapid tests in general will grow not more than 4% annually for the foreseeable future.

2.2.7 Lateral Flow Tests in Clinical Diagnostics

2.2.7.1 Clinical Test Segments

This biggest LF test segment is generally seen as the most lucrative because, worldwide, there is a huge demand for decentralized availability of diagnostic tests. Clinical LF-based rapid testing generated company revenues of $2,010 million; and accounts for 89% of the total world market for lateral flow tests in 2007 (Table 2.4). With an 10% CAGR, this segment will grow to $3,240 million by 2012.

Clinical LF tests are used primarily to replace lab-based immunoassays in decentralized testing locations, generally known as point-of-care testing (POC). The data presented here include all LF-based POC immunoassays for self-testing, which are sold over the counter (OTC) and those performed by health-care professionals in hospital laboratories, hospital wards, clinics, community health centers, and physician offices. The data do not include the diabetes segment. These tests do not use LF technology. LF tests are sometimes used in laboratories when a rapid test result is required. However, the success of LF-

Table 2.4 Clinical POC sales by test category, worldwide, 2007–2012 (in $ Millions)

	Sales 2007	% Market	Sales 2012	% Market	% CAGR
OTC & Professional					
Pregnancy	695	35	775	24	2
Infectious Diseases	400	20	850	26	16
Cardiac Markers	425	21	850	26	15
Cholesterol/Lipids	260	13	490	15	14
Drugs of Abuse	145	7	165	5	3
Other	85	4	110	3	5
Clinical Total	2,010	100	3,240	100	10

Other = TSH, PSA, FSH, HbA1c, cancer, etc.
Source: Kalorama Information, company reports

based tests is primarily linked to trends in POC testing, the venue for which they were originally designed.

POC testing, by lateral flow, inside and outside the hospital, is one of the fastest growing segments of the diagnostics market. In 2007, LF tests represent approximately 35% of the immunoassay market, compared to 25% in 2003. This growth is largely related to significant increases in OTC pregnancy tests, and in various professional test segments including cholesterol, cardiac markers, and others as shown in Table 2.4.

Since the early 1990s, industry watchdogs predicted that POC would be the dominant methodology used when rapid diagnoses are required. The prediction has not been realized. Infrastructure and reimbursement barriers have been blamed for POC testing's poor track record. However, worldwide, payer groups have begun to provide reimbursement approval for POC tests. Many physicians use POC testing as a tool to provide improved patient care. Inside the hospital, the problems associated with quality control and training non-laboratory staff have scared off many would be users. To avoid these issues, many hospitals have placed rapid test systems in laboratories, or set up special POC test teams.

Therefore, with many administrative issues under control, the most telling cause for POC's limited market penetration is really technological. Most rapid POC tests are based on immunochromatographic techniques with visual detection of test results. However, as imaging and computational capabilities improve, there is a significant trend within the diagnostics industry to replace visual detection with digital and instrument-based methods.

Most new POC test devices have a built-in memory to store test results that can later be downloaded into a computer or have the ability to send results to a receiving network. Further, they are being designed with multi-test cassettes that avoid the need for quality control procedures each time a test is run. Improvements in all of these aspects of POC testing have gone a long way to increase testing in the professional sector.

The next step is to make home testing for chronic diseases interactive so that patients can be advised to take action when necessary. Similarly, POC information systems can aid physicians in determining the significance of a lab result. This value-added component of POC testing is in progress.

The most widely used home (over the counter, OTC) LF tests are for pregnancy and ovulation. Worldwide, these home tests generated approximately $565 million in 2005; 87% of the market for all OTC tests. Home-based tests for drug use have had a limited appeal to parents primarily in the United States. Other tests such as those for cholesterol, infectious diseases, fecal occult blood and coagulation have not had the anticipated market penetration. Manufacturers had anticipated consumer acceptance of these and other self-tests. The need for a blood sample is a primary reason for the limited use of these tests.

Added to this is a significant dilemma for the users of self-tests – what does it all mean? Consumers are often confused about what to do with the test results and so prefer just not to do them. The time is ripe for more user-friendly digital OTC devices that leave nothing to interpretation. Launched in 2004 are OTC pregnancy test devices that say yes or no instead of a series of lines or dots. The public is ready. Now it is up to the manufacturers. In the meantime, the latest trend to overcome reticence for self-testing is direct to consumer marketing of lab tests in public venues such as malls and supermarkets and home testing kits via the Internet.

After many years of slow growth, the professional POC test market is beginning to come alive. Demand for quicker test turnaround time has spurred the launch of at least 60 new POC tests and devices in the past few years and at least another 30 are near market. The market for professional LF POC tests is estimated at $1,375 million up from approximately $810 million in 2000. With 7% growth, this market segment will reach $1,950 million in 2012. Most of the growth will come from increased use of cardiac markers and tests for critical care. Other growth stems from new assays for cancer, diabetes, cardiac disease, lipids, and coagulation factors.

Not all POC testing needs can be filled by LF-type rapid tests. Increasingly, physicians are looking to increase the number of tests they offer in their offices. There is growing demand for small, easy-to-use, and easy-to-maintain systems for near-patient sites and small clinical labs worldwide.

In the hospital setting, there is pressure for more rapid turnaround time and efficient patient management to minimize the length of stay in emergency rooms and throughout the hospital. However, hospital administrators maintain that POC tests (LF included) are more expensive than lab-based tests and that manually read patient test results are lost to the historical record. Further, once the patient leaves an acute care area such as the ER, OR, or an intensive care unit, the baseline testing done in that unit is relatively useless because, more often than not, the test results from LF devices do not correlate well with lab-based systems. The evidence for this supply/demand dynamic is evident in the continued growth in the market for critical care analyzers and instrumented systems for cardiac and cancer markers.

Outside the hospital, the most widely used professional POC tests are for pregnancy, infectious diseases, cholesterol, and urinalysis strips. In Europe, numerous companies market small tabletop chemistry and immunoassay systems for physician office labs (POLs) and clinics. In the United States, this trend is emerging. For a number of years, POLs had resorted to performing only Clinical Laboratory Improvement Amendment (CLIA) waived tests. However, in the past several years, companies have begun distributing chemistry, hematology, and immunoassay instrumentation to POLs more aggressively. These compete significantly with LF tests especially in the areas of infectious diseases, cardiac markers, cholesterol, and hormone tests.

There was some anticipation that rapid tests for *H. pylori*, Strep A, flu, cholesterol, Lyme diseases, and sexually transmitted disease (STD) (e.g., chlamydia and vaginitis) would become part of physicians' care routine. This has not been the case, although at least 75 vendors worldwide market POC test kits for these analytes and more. There are a number of reasons to account for the limited use of these tests worldwide. Firstly, the tests take too long – very few patients will wait for approximately 30 min for test results after they may have already spent at least that long waiting for their turn to see the physician. Further, technologically, many of the tests have not proved to be as sensitive as lab-based tests.

Here the future use of miniaturized lab-on-a-chip type devices should make rapid, cost-effective assays for just about any assay available for POL use. These are not expected to make any market impact until 2015.

The LF market is highly diversified, with a different set of companies active in the various sectors. At least 20 companies are active in the largest self-test segment – pregnancy and ovulation (LH). The top vendors are: Inverness Medical/Unipath (Waltham, MA), Carter Products/Division of Carter-Wallace, Inc. (New York, NY), Parke-Davis/Warner Wellcome Consumer Health (Morris Plains, NJ), and Ortho Pharmaceutical Corp. (Titusville, NJ). Further, all major pharmacy chains sell their own home brand test kits manufactured by the leaders and other original equipment manufacturer (OEM) companies. This test segment continues to grow at an average of 2% per year as more and more women enjoy using these tests in the privacy of their own homes.

Roche Diagnostics (Indianapolis, IN) pioneered the coagulation self-test market since the mid-1990s and has had some success in Europe, primarily Germany. However, without reimbursement, there was little demand in the United States. Then, in July 2002, the US Medicare program began to cover home anticoagulation monitoring for patients with mechanical heart valves that are taking coumadin. This segment should therefore see 11% annual growth. In the past several years, several more POC pregnancy tests (PT) have come to market, most are used by professionals in the hospital or clinic setting. Many pharmacies maintain CLIA labs and are already preparing to offer PT testing to the patients unable to perform the tests themselves. Payment can be arranged with health insurance providers for qualified patients or would be paid out-of-pocket.

Most professional cholesterol testing takes place in outside the hospital in patient care sites such as physician offices, clinics, and wellness fairs. The market is growing a healthy 14% annually and is estimated at $260 million in 2007. The market leaders are Roche's Accu-Chek and Reflotron products, and Inverness/Cholestech's LDX. Other players are ActiMed Laboratories' (Burlington, Nj) ENA.C.T. Total Cholesterol Test, Johnson & Johnson's ADVANCED CARE Cholesterol Test, and Polymer Technology Systems, Inc.'s (Indianapolis, IN) Bioscanner Test Strips.

As more is learned about cardiovascular diseases and their links to diabetes, lipid metabolism, inflammation, and hypercoagulopathies, the definition of a cardiac marker expands. Depending on one's point of view, tests such as D-dimer, hsCRP (high-sensitivity C reactive protein), HgbA1c, cholesterol and associated lipid fractions can all be considered part of the cardiac marker segment. Thus, this is by far the most dynamic POC segment that will show the most change over the next few years, particularly in the assembly of multi-analyte panels that draw from all of these areas. In the meantime, traditional cardiac markers (CK-MB, myoglobin and troponin I/T) have had their own shake up with the emergence of D-dimer, hsCRP, and BNP (B-type natriuretic peptide) as new additions. The undisputed leader and innovator in this segment is Biosite, Inc. (now part of Inverness Medical). There are also at least 20 companies that sell LF tests for all or one of the cardiac marker panels.

At least 75 companies sell rapid tests for infectious diseases including Strep A/B, legionella, HBsAg test, mononucleosis, malaria, respiratory viruses, influenza, meningitis, filariasis, adenovirus, tuberculosis, Epstein Barr virus, measles, mumps, rubella, chlamydia, and gonorrhea. The primary market for LF-based infectious disease testing is in patient care settings outside the hospital. The market leader is Inverness Medical with 35% of the market. Then comes Becton Dickinson and Company (Franklin Lakes, NJ) with 15%, Meridian Bioscience, Inc. (Cincinnati, OH) and Quidel Corporation (San Diego, CA) with about 8% each, and after that a host of some 20 companies including: Inverness/Acon Labs, Beckman Coulter (Fullerton, CA), Fisher Healthcare (Houston, TX), Genzyme/Wyntek (Cambridge, MA), Gull Labs, Polymedco, Inc. (Chicago, IL), Princeton BioMediTech Corporation (Princeton, NJ), Trinity Diagnostics PLC (Bray, Ireland), and Remel (Fisher Scientific). This sector will see a moderate 16% CAGR due to increased testing for respiratory disease pathogens. However, rapid immunoassays are threatened by breakthrough in molecular testing with increases in sensitivity and specificity of these tests.

Most cancer testing is done by lab-based immunoassays, molecular tests, and analysis of tissue biopsies. The newest LF test on the market is Inverness/Matritech's BladderChek Test POC test for bladder cancer. As of June 2006, Matritech reports that more than one million BladderChek Tests were sold. The product accounts for approximately 75% of the company's product sales ($10 million).

2.2.7.2 Clinical Test Geography

The US market accounts for $1,005 million (50%) and the European market for approximately $804 million (40%) of the worldwide market for LF clinical tests. Japan and Asia represents 5% ($100 million) and the ROW accounts for the remaining 5% of the market. As the largest single market for in vitro diagnostic (IVD) and POC products, the United States has an enormous impact on how the rapid test industry develops. Managed care's obsession with cost reductions is pushing the need for nearer the patient and decentralized testing in the home, at the bedside and in the physician's office. This implies a level of connectivity not generally available in current POC devices but which is becoming a necessary feature in new product design.

The European market for POC tests, especially in the POL sector, has been growing faster than that of the United States. However, most of these test locations use low volume, tabletop versions of traditional chemistry, immunoassay and hematology instruments versus specially designed POC devices. The main reason for this is that these laboratories are lightly regulated compared to the CLIA standards required in the United States. Therefore, the European market for LF-based devices has been evolving slower than that of the United States. This situation may change as EU governments see centralized testing as a way to control the cost of delivering healthcare to their aging populations. Similar to all countries of the OECD (Organization for Economic Co-operation and Development), they have begun placing greater emphasis on the prediction and prevention of disease through more proactive diagnostics. Thus the POC test market in Europe is expected to increase in the number of tests conducted, the number of tests available, and the number of test locations.

Japan's highly centralized form of healthcare delivery does not leave much room for POC testing in the physician office. However, Japan's hospitals are increasing their uptake of rapid tests for emergency and critical care. Further, Japan is the single largest market for rapid flu tests since Roche's Tamiflu is prescribed for all diagnosed cases of influenza.

The combination of increasing international demand for sophisticated laboratory systems and quality rapid test products, in conjunction with the slowing of IVD market growth in OECD countries, has encouraged companies to look outside the traditional markets – the United States, Europe and Japan. Market opportunities are evolving in emerging markets such as South America, Eastern Europe, Russia, and parts of Asia and Africa. Entrepreneurial companies are beginning to take advantage of them, more as a method of survival than for altruistic reasons.

2.3 The Future

The roots of lateral flow technology go back to the discovery of the antibody–antigen immunoassay reaction by Yallow and Burson in the 1960s combined with the principles of thin layer and paper chromatography. Then in 1987, and within several months of each other, three researchers, Robert Rosenstein for Becton Dickinson & Co., Keith May for Unilever, and David Charlton for Carter Wallace, filed US patents for what is now considered the basics of the LF platform. BD still owns the Rosenstein patent and Inverness Medical acquired the other two patents.

In an interview with Robert Rosenstein, he explained that all three patents did essentially the same thing with slight variations. The aim was to develop an easy OTC platform for the pregnancy test. "Because of the significant implications of these almost simultaneous filings, it took 10 years for the patents to be issued in 1997. This ten year delay impeded innovation at a time when the demand for rapid tests was expanding, The result is that, up until recently, all LF tests are just about the same – same analytes and same technology," he added.

How did the patent delay impede innovation in rapid tests? A patent does not prevent competition, it encourages competition. Once the patent is published, the whole world knows what the patentee is doing and has a detailed road map of the invention and its intended use. In this way it leaves the door open for other inventors to change or improve upon the original patent and thus allows for some competition in the time frame of the original patent.

The three base patents and add-ons are set to expire in 2006–2013. This has inspired a wealth of innovation that should put LF in good stead for the next 5–10 years or until digital multiplexed platforms gain market penetration.

"Lateral flow is limited, so companies are starting to develop readers to remove the subjectivity in reading the test strips. It also makes lateral flow not such a simple test to use because readers have to be calibrated. There are ways to adapt lateral flow to new test needs, but it won't happen for every kind of test. Lateral flow is best for yes/no type tests. If you need quantitation, then they probably are not as good as some of the other new technologies are more appropriate. So lateral flow will definitely be complemented by other testing modalities such as chips, biosensors and bead arrays," said Rosenstein.

A small example of the innovation in new tests and techniques underway is shown below.

1. BioAssay Works is developing a new colloidal gold that minimizes background noise and enhances the reaction.
2. Chembio Diagnostic Systems has received a patent for its next-generation Dual Path Platform (DPP), a chromatographic immunoassay platform that Chembio believes improves sensitivity in single and multiple parameter tests as compared to standard single-path lateral flow assays.
3. Cibitest GmbH & Co. KG is developing its FLORIDA technology (Fluorescence Labelled Optical-Read Immuno Dipstick Assay) designed for lateral flow immunoassays with a sensitivity of only a few parts per trillion.
4. EY Laboratories has developed the InstantCHEK that provides quantitative test results for infectious diseases in seconds.
5. LifeAssays AB has developed the Magneto-Immunoassays in a lateral flow format, which replace the gold or colored polymer particles with superparamagnetic particles as the label.
6. QuantRx Biomedical was issued a US patent for the QuantRx oral fluid collection device specifically designed for rapid lateral flow. It allows for the production of a one-step device incorporating saliva sampling and testing.
7. Talecris Biotherapeutics and BBI Holdings PLC are developing a physician office test device as a screening tool for detecting patients' levels of Alpha-1 Antitrypsin (AAT), the naturally occurring protein that is present at low levels in individuals suffering from Alpha-1 Antitrypsin Deficiency.
8. The UK's Institute of Biomedical Science is promoting the development of particle-based immunoassays incorporating lateral flow membrane technology for a rapid allergy assay using a 20-μl sample of whole blood.
9. WaveSense LLC introduced the UltraPlatform, a proprietary magnetic bead–based lateral flow separations system, which is designed to capture and sort live cells, proteins, and or genes for testing.

References

1. The LF data for this chapter is taken with permission from The Worldwide Market for In Vitro Diagnostic Tests, 5th Edition April 2006, www.kaloramainformation.com and other reports authored by Shara Rosen and published by Kalorama Information, New York, NY.
2. Guidance for Industry, PAT – A Framework for Innovative Pharmaceutical Development, Manufacturing, and Quality Assurance, U.S. Department of Health and Human Services, Food and Drug Administration, Center for Drug Evaluation and Research (CDER), Center for Veterinary Medicine (CVM), Office of Regulatory Affairs (ORA), Pharmaceutical CGMPs. September 2004, www.fda.gov/Cder/guidance/6419fnl.pdf

Chapter 3
Material Platform for the Assembly of Lateral Flow Immunoassay Test Strips

Jennifer S. Ponti

3.1 Introduction

Lateral flow immunoassays provide a convenient and relatively inexpensive means of performing biological and chemical testing. The technology is used to support existing and emerging markets and applications including:

Infectious Diseases	Clinical Diagnostics
Drugs of Abuse	Forensic Science
Human/Animal Health	Therapeutic Monitoring
Genetic/Cardiac Markers	Agriculture
Environmental Safety	Water/Food/Dairy Safety/Processing
Bio-Defense	Biotech Quality Control

And with many more uses being discovered daily, this exciting technology has almost unlimited potential.

Material selection is a critical criterion for the success of any lateral flow platform. A test strip is typically composed of a plastic backing with an adhesive, a sample pad, a conjugate pad, a nitrocellulose membrane, an absorbent pad, and printed tapes for identification. This chapter will outline the materials that form the backbone or platform of a testing device including backing basics, adhesives, liners, backing format, sub-assembly concepts, cover tapes, and laminating and processing recommendations.

3.2 Selection and Assembly of Materials Used in Test Strips

3.2.1 Backing Basics

Many of the components of a lateral flow immunoassay test strip, such as nitrocellulose, fiberglass pads, and cellulose sample and absorbent pads, are

J.S. Ponti (✉)
G&L Precision Die Cutting, Inc., San Jose, CA
e-mail: jponti@glprecision.com

R.C. Wong, H.Y. Tse (eds.), *Lateral Flow Immunoassay*,
DOI 10.1007/978-1-59745-240-3_3, © Humana Press, New York, NY 2009

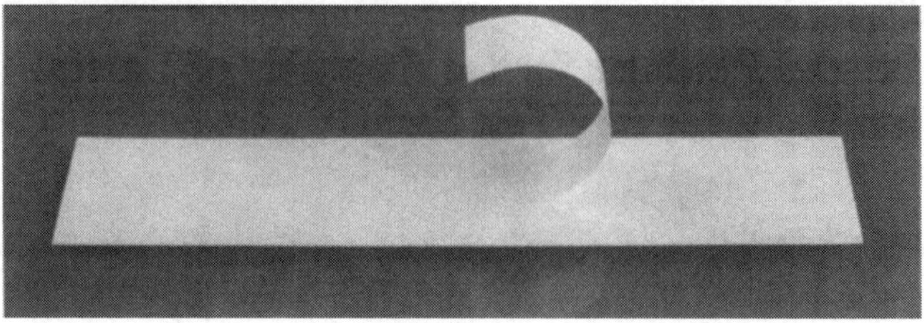

Fig. 3.1 A typical backing with part of release liner partially peeled back. (Photo courtesy of G&L Precision Die Cutting, Inc. San Jose, CA, USA)

delicate in nature. Therefore, test strips need a supporting backbone or platform to run on. This platform is commonly referred to as the backing or backing card as shown in Fig. 3.1.

The backing is designed as a custom platform to fit the specific requirements of each test. A traditional backing consists of three components – a semi-rigid plastic, a pressure-sensitive adhesive, and a release liner. The most common plastics used for lateral flow immunoassay tests are polystyrene, vinyl (poly vinyl chloride or PVC), and polyester. Custom materials are also available from experienced converters and in vitro diagnostic (IVD) component suppliers like G&L Precision Die Cutting (San Jose, California), Millipore, Inc. (Billerica, Massachusetts), Adhesive Research, Inc. (Glen Rock, Pennsylvania), and Whatman Inc. (Florham Park, New Jersey). Commonly available thicknesses are:

a) 0.005 inch for polyester
b) 0.010 inch for polystyrene, polyester, and vinyl
c) 0.015 inch for polystyrene and vinyl

The thickness of the backing material is a major consideration for the design of a test strip. Thinner backings in the 0.005 to 0.010 inch thickness are typically used for test strips that go into a holder, cup, or plastic housing. Films in the 0.010 to 0.015 inch range and thicker materials are usually preferred for stand-alone test strips. Backing materials greater than 0.015 inch in thickness should only be used if absolutely necessary since thicker backing materials can cause difficulty in assembly because of the propensity of the materials to curl. It can also create issues during cutting of the material into individual test strips. Thicker materials will require greater cutting force that can cause ridge marks along the edge of the strip or twist the strip, which can affect sample flow. In general, the backing material should be thick enough to provide support to the strip during manufacturing and assembly, but not too thick that it creates irregularities when cutting assembled material to the finished test strip size, or adds unnecessary cost to the product.

Custom films, absorbent materials, and rigid plastic are also used sometimes as backing materials for very special applications. In initial product design, one may wish to research or consult with one or more experienced converters to best determine the materials that are available for use in the new product.

3.2.2 Backing Format

Plastic backings are available in pre-cut card form (or called sheets) and roll form. Cards are the most common and flexible format for many manufacturers, but there is an increasing demand for backing in roll form for use on automated or semi-automated equipment. Backing materials have a propensity to curl, causing liners and adhesives (see below) to pinch or wrinkle if not handled properly. Precautions should be taken in determining the format that will actually work best for a specific application. There are some general guidelines to consider for the two formats:

a) Backing cards allow for a wide range of manual assembly options. They are easy to handle and promote flatness of the material. They can also be partially assembled and easily stored for future use or finishing. Cards can be a very effective format and are still the most popular format today. However, cards cannot be easily automated and will generally be labor-intensive to assemble.
b) Backing in roll form can be used on an in-line automated assembler for high-volume production. However, because lateral flow laminations (see below) generally result in a fairly thick laminate with varying heights, it is not recommended that the materials be rewound after being laminated. Therefore, the laminated materials need to be cut either into cards or into test strips in-line. If the latter case is chosen, a system should be designed so that the laminating portion can still be done at high speeds.

3.2.3 Diagnostic-Grade Adhesives

One of the most important issues to consider in choosing a backing for a lateral flow application is the adhesive used on the backing material. Not all adhesives are created equal and given the expense of most other lateral flow components; this is not an area to cut costs on. Special attention should be made to choose an adhesive with the following characteristics:

- Does not build up excessively on blades during cutting
- Does not migrate into conjugate pads, sample pads, or backed or unbacked membranes (i.e., an inert adhesive)
- Compatible with biological components of the test and does not affect the stability of reagents or membranes (especially for acrylic adhesives, see below)
- Compatible with a stable protective release liner
- Specially formulated diagnostic-grade pressure-sensitive adhesive

It has been shown that rubber and industrial acrylic adhesives can leach chemicals, solvents, and other additives. This property may affect the bonding of materials, or the chemistry of the test, causing false positive or false negative results. Solvents and plasticizers, common in acrylic adhesive systems, are especially prone to migrate out of the adhesive. Users should always seek out a diagnostic-grade adhesive from an experienced supplier or converter to avoid such complications.

Pressure-sensitive adhesives, commonly referred to as "psa", are used in most lateral flow applications. These adhesives form a bond when pressure is applied. No heat, water, or other outside chemicals are required. Typically, a psa will "wet out" or cure and achieve its ultimate bond over the course of 24 hours. This is important when using substrates that are difficult to bond to, such as a fiberglass pad. If components are not bonded well or are falling off during the process of cutting the assembled materials into test strips, the cause could be that not enough time has been allowed for the materials to form a sufficient ultimate bond.

3.2.4 Liners

In order to protect the adhesive before and during assembly, a protective liner must be placed over the adhesive. The liner can be a paper or a film treated for release characteristics. A paper liner must be further treated for moisture resistance. The release agent is typically a silicone coating to allow for easy removal of the liner from the adhesive.

For in vitro diagnostic (IVD) applications, the liner is generally scored with "backsplits" to aid in device assembly. A backsplit is a custom, pre-cut score in the liner, sometimes referred to as a kiss cut, backsplit, or back score. A proper backsplit will be cut completely through the liner, just barely touching the adhesive. A backsplit should never be cut through the plastic backing as this jeopardizes the integrity of the material. Figure 3.2 depicts an example of the

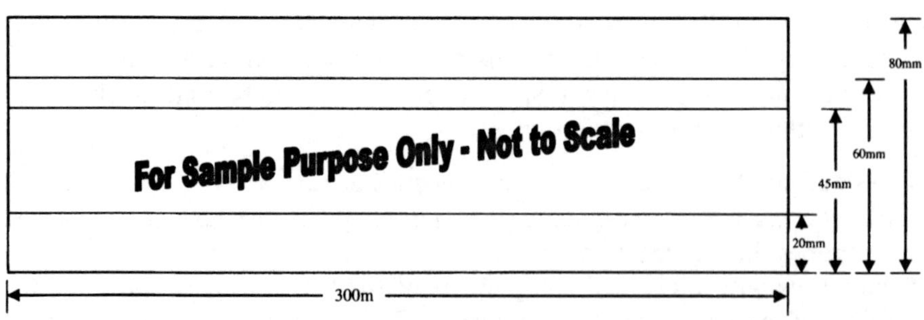

Fig. 3.2 Dimensions for a typical lateral flow immunoassay test strip (for sample purpose only, not drawn to scale)

dimensions for a typical lateral flow immunoassay test. Note that all dimensions are measured from the bottom.

The various components from bottom to top are: the absorbent pad (20 mm), the nitrocellulose membrane (25 mm), the conjugate pad (15 mm), and the sample pad (20 mm). A custom cutting die for scoring the backsplit, often referred to as "tooling", is made for each unique product configuration. This is a one-time-only investment, as the tool can be used over and over again.

Determining the locations of all of the components in a lateral flow immunoassay test including overlaps is critical to the success and reproducibility of a test strip. An important item that should not be overlooked in determining the overlaps and backsplit locations is the tolerance on the widths of the materials. Typical tolerances on the width of the sample pads, conjugate pads, membrane, and absorbent pads are in the ±0.25 mm to ±0.50 mm range and will vary by manufacturers. Typical tolerances for the width of the backing are ±0.35 mm and those of the backsplit locations are ±0.005 mm. It is important for the success of a test that the overlaps are correctly defined in a way compatible with the dimensions of the test strip and taken into consideration the limits of the tolerance.

3.2.5 Sub-assembly and Contract Manufacturing

Experienced converters and suppliers often provide a number of value-added features and sub-assembly options to help customers get their products to the market faster. Sub-assembly services may include custom lamination of customer-supplied materials such as nitrocellulose membrane, conjugate pads, absorbent pads, and cover tapes. They may also offer the assembly of active materials in dry rooms. Close tolerance laminating of components often translates into improved consistency and accuracy of a lateral flow immunoassay test. It can also reduce the high-waste factors typically associated with hand lamination. Overall, it decreases costs by allowing a manufacturer to eliminate or reallocate extra labor and to handle production spikes in volume without adding staff. In addition, it may free up time to let researchers focus their development efforts on chemistry and design of the test strip and ultimately get their products to the market faster.

In addition to the sub-assembly type services, converter partners may offer custom options like notching the backing cards for registration, improving placement of components, zone coating of adhesives, and producing adhesive free areas on the card.

As the market moves toward quantitative tests or improved sensitivity of tests, many manufacturers search for better tolerances on the width of raw materials. An experienced converter partner can provide narrow web slitting of lateral flow components often to a closer tolerance than available from the raw material supplier.

3.2.6 Cover Tapes

Printed films are used as a top laminate or as a cover tape in many lateral flow applications. Typical uses for cover tapes include:

a) Test identification
b) Arrows to indicate orientation of the test strip
c) Stop lines to indicate the upper limit in which the strip should be immersed into the liquid sample
d) Trade names
e) Identification during manufacturing
f) Holding down fragile components (i.e., fiberglass pads)

Cover tapes materials are generally thin, flexible film tapes with adhesive on one side and printing on the other. The film can be either clear or opaque white and a Flexographic printing process is used to print virtually any color, text, or symbol on it. A unique print plate is needed for each design. Printing can be positive print where the ink is the color of the text, or reverse print where the text is in white surrounded by color.

Gaining popularity is the practice of printing cover tapes in different colors for each product a company manufactures. This allows easy identification of the actual test during manufacturing when several test strips on the same platform are similar in size and base materials, but with different chemistry. Often, the finished test strips are assembled into housing, which does not show the identification tape in the final packaging.

Cover tapes need to be evaluated for performance whenever a change is made to a test including materials, formulation, and test strip cutting. For example, a change from a cellulose-based pad to a glass fiber–based pad may significantly improve the performance of the test strip, but the cover tape may not be aggressive enough to bond to the new fiber glass pad that has a very open cell structure. Another common challenge occurs when the method of cutting test strips is changed. Going from a guillotine test strip cutter to a rotary cutter may improve manufacturing efficiencies. However, since many male/female rotary test strip cutters twist the strip during the cutting process, this may cause the cover tape to pull away from the taping components or even delaminate entirely from the strip.

3.2.7 Assembly and Laminating

The assembly of a lateral flow immunoassay test is as varied as the tests themselves. There are hand assembly, hand-made jigs, light box, vacuum shell systems, large, automatic in-line lamination systems, and everything in between. All methods are effective and can be adapted to meet a variety of manufacturing requirements.

Hand assembly is sometimes designed so that only the backsplits in the liner guide the assembler to where to put the components. A jig or light box will show lines indicating where each component goes and aid in manual hand assembly.

Several companies offer clam-shell style laminators that use pin locators, formed trays, and vacuum pressure. The components are placed on one side with the proper overlaps and then the clam shell is closed, pressing the two sides together and laminating the materials onto the backing card. In addition to these kinds of batch laminators, equipment manufacturers also make custom in-line lamination systems. A properly designed in-line system provides individual control to exert proper tension to the various components that make up the lateral flow immunoassay test.

The wide variety of equipment and design support allows researchers to scale-up quickly from bench, through pilot production to full manufacturing with a minimum of process development issues.

Regardless of the assembly method, whenever possible, time should be allowed for all of the materials to form an adequate bond once laminated. A period of a few hours will suffice for many tests, but some tests with high surfactant loads or fiberglass pads or others difficult to bond to substrates may require a longer period of time.

3.3 Conclusion

In the development of lateral flow immunoassay, most attention has been paid to finding a good detection method or seeking the best antibody or antigen. However, as we have shown above, one should ensure the compatibility of basic components like the backing, adhesive, and cover tape in order to produce a consistent and high-quality product.

Chapter 4
Antibodies: Key to a Robust Lateral Flow Immunoassay

Michael C. Brown

4.1 Introduction

Lateral flow immunoassays are well established as a valuable tool in medical, veterinary, food, agricultural, environmental, and industrial diagnostics. While in many instances they are used as a rapid screening tool and backed up by more complex and time-consuming assays, they may also be used in some applications as the primary method of choice. As a consequence, these assays need to be constructed as robustly as any other diagnostic tests. While the physical components of the lateral flow test strip, construction techniques, and buffers and surfactants play a major role in optimizing the test, at the very heart of these processes are the antibodies themselves. If the antibodies are not meticulously selected for their ability to recognize the target antigen(s) in this format, no amount of optimization will be able to overcome this inherent defect. Often, assay developers may spend great deal of time struggling to "tweak" the assay to fit their specifications, but come away with an assay that just borders on having the right sensitivity and generally not very robust from a manufacturing standpoint. Instead, if they would spend more time in the creation and selection of optimized antibodies at the front end of the development process, it would make the rest of the development relatively easy.

Lateral flow immunoassays are particularly demanding in terms of the affinity required in the interactions between antibodies and antigens. Consider a typical lateral flow strip with antibody immobilized on a test zone of 0.5–1.0 mm wide. Antigen flowing up the strip has a flow rate in the range of 0.66–0.16 mm per second depending on the flow rate of the nitrocellulose membrane selected (most often specified by the manufacturer as the time required to traverse 4 cm of the membrane and it ranges from 60 to 240 seconds [1]). Antigen thus spends between 1 and 6 seconds in the zone where it can interact with the immobilized antibody. The speed is actually much faster at the initiation of the flow, since flow rates decrease proportionately to the square of

M.C. Brown (✉)
Strategic Diagnostics, Newark, DE
e-mail: mbrown@sdix.com

R.C. Wong, H.Y. Tse (eds.), *Lateral Flow Immunoassay*,
DOI 10.1007/978-1-59745-240-3_4, © Humana Press, New York, NY 2009

the distance traveled until they achieve a steady flow rate when the entire bed volume of the nitrocellulose becomes saturated. Furthermore, most strips employ nitrocellulose path lengths much shorter than 4 cm, so their flow rates are significantly faster than the 60–240 seconds per 4 cm used in the above calculation. The situation is somewhat better for antibodies that are coupled, or conjugated, to detector particles, such as colloidal gold or latex. These reagents are often dried into a pad at the base of the nitrocellulose and solubilized by the initiation of liquid sample flow (see Chapter 1). For a conjugated particle initially released, its time of interaction with antigen is now the time from exposure to antigen and solubilization to the time it passes through the test zone. This could be as few as 8 seconds for particles at the leading edge or as long as the entire duration of the test for a small number of particles that release and move slowly. Thus the requirements for high affinity (and hence short speed of reaction) apply to all antibodies used in lateral flow immunoassays, regardless of where in the strip they are used.

The mass of the reagents used in a lateral flow immunoassay test also plays a significant role in driving the antibody/antigen interaction. When antibodies are used in a test line in a sandwich-type assay, they are applied at a ratio of 1–3 μg per cm across the width of the nitrocellulose strip in a line 1 mm wide and a relatively shallow bed volume of 0.13 mm. This results in an antibody concentration of 10–30 μg per square cm, which is 25–100 times that used in ELISA at a typical maximum concentration of 300 ng per square cm [2]. Furthermore, the capillary flow through the bed of the membrane forces proximity of the reagents, rendering the distance required for diffusion to be proportionately very small when compared to the pore size, which is in the range of 1–10 μm.

In lateral flow immunoassay test strips, the most utilized size of colloidal gold antibody particle is 40 nm in diameter [3]. Each test strip would typically use 0.03–0.25 μg of the colloidal gold antibody. On hydration, assuming volumes of 10–50 μl at the zone of maximum concentration during flow initiation, this would result in effective antibody concentrations of 0.6–20 μg per ml. While these relatively high densities of antibodies drive the antibody–antigen interaction by mass action, they can also contribute to non-specific binding and lack of specificity or cross-reactivity in some instances. Careful selection of antibodies can allow a density that supports the desired sensitivity but does not contribute to non-specific interactions. Far too often, a developer will attempt to gain additional sensitivity by putting excess reagents into the system and the frequent result is deterioration of the assay due to non-specific interactions.

In summary, the development of robust lateral flow immunoassay tests requires antibodies that are highly specific and have high affinity for the target antigen. No amount of development effort and optimization can make up for having selected the wrong antibodies at the beginning of development. To this end, this chapter will discuss the sources and selection of antibodies, the relative advantages and disadvantages of monoclonal and polyclonal antibodies, the role of genetic immunization, antibody purification, and the testing and quality control required to deliver a robust set of reagents and methods to manufacturing in lateral flow immunoassay test development.

4.2 Commercial Sourses of Antibodies

Antibodies suitable for lateral flow immunoassays are available from a number of commercial sources [4]. Frequently, these antibodies can be obtained prequalified by the vendor in pairs suitable for sandwich assays. These pairs are most readily available for relatively common and high-volume assays, such as tests for pregnancy (human chorionic gonadotropin), infectious disease (HIV, hepatitis B, *H. pylori*), cardiac markers, (tropoinin C, creatinine kinase- MB, myoglobin), and malignancies (prostate-specific antigen). Similarly, antibodies suitable for use in competitive assays, such as hormones, therapeutic drugs, and drugs of abuse, are available. It is also possible to obtain some of these antibodies already conjugated to colloidal gold or latex particles. For novices to lateral flow immunoassays, purchasing of a small amount of these antibodies represents a good initial investment as a learning tool. Using antibodies and other components of known performance, one can standardize and configure specific lateral flow assays. If the interest is in targeting the common analytes, purchasing these reagents is a quick way to get to market, but assay performance will likely be comparable to the tests already on the market and commodity pricing may be the key to success. One should approach using commercially available antibodies by asking the following questions:

i) Do the vendors actually produce the antibody or are they simply brokering it?
ii) How is this antibody quality-controlled?
iii) Can the vendors guarantee supply in the long term?

If the answers to these questions are not crystal clear, one should proceed with caution. There are a great number of antibodies sold by multiple vendors that in fact have a single source. If the developer can buy directly from that source, many of the supply chain problems that commonly occur down the road can be averted.

Purchase of "off the shelf" antibodies may be considerably more expensive on a milligram basis than the antibodies custom-generated and owned by the end user. The trade off, of course, is that custom antibody development costs are eliminated, or at least those costs are not readily apparent. Given the large amounts of antibodies consumed in commercially successful assays, pricing may play a significant role. At $100 per mg, a typical assay could add $0.05–$0.07 in antibody costs alone. Generating one's own antibodies or contracting with a commercial source for antibody generation may, in the long term, provide a price advantage as well as the opportunity to create unique characteristics of the resultant immunoassay.

4.3 Selection of Antibodies

Careful selection of antigen and its purity for the generation of high-quality antibodies is of paramount importance. Impurities will give rise to poor antibody performance and contribute to reaction that may be difficult, though not impossible, to sort out. One exception is the generation of monoclonal antibodies with a well-planned screening methodology. Indeed, monoclonal antibodies can be raised quite successfully using extremely crude antigen when rigorous screening and selection strategies are in place. Molecules smaller than 6,000 kD usually require an assay format based on a competitive reaction. In the simplest form of a competitive assay, target antigen competes with a labeled form of the antigen for antibody binding. This assay typically signals binding when the competing antigen is not present and shows reduced binding when the competing antigen is added. Such a format is used with antigens of relatively small sizes because the physical space for binding of multiple antibodies would result in steric hindrance.

Antigens of smaller sizes (e.g., a hapten), particularly those smaller than 1 kD, also do not make optimal immunogens. They may rapidly diffuse from the site of injection and be cleared from the system by mechanisms other than the immune system. In order to elicit an immune response to small molecules, these antigens are coupled to a carrier proteins [5], often keyhole limpet hemocyanin (KLH) or ovalbumin (OVA). An optimal carrier can be sparingly soluble, and remains at the injection site to give a prolonged reaction with cells of the immune system. It may also be strongly antigenic, capable of provoking a strong immune response. Keyhole limpet hemocyanin [6] has both of these attributes. Furthermore, the carrier used should not be naturally present in testing samples so as to avoid having to distinguish antibodies to the carrier from the antibodies to the hapten. In this regard, one might be advised against using serum albumen as a carrier for immunization. The carrier should also have a plurality of functional groups (amines, hydroxyls, carboxyls, or sulfhydral groups) to which the antigen may be coupled.

To couple the antigen to a carrier, a crosslinking agent is required to covalently link the two molecules together [7]. It is often advantageous to include in this crosslinker a spacer arm to position the antigen away from the surface of the carrier for optimal presentation. Homobifunctional crosslinkers have two identical reactive groups and can result in significant crosslinking and polymerization. These are usually used in a single-step process. Heterobifunctional crosslinkers have two different reactive groups. This allows a more sophisticated crosslinking strategy, often in several steps. Typical for many haptens, an activated derivative is first prepared. Amino, hydroxyl, or carboxyl groups at positions not critical for distinct characteristics of the hapten may be converted, through a series of simple reactions, to succinimidyl esters [8]. These activated compounds are generally quite stable for years when kept dry. At pH 7.5–9.0, they will couple quantitatively to the carboxyl or amino groups of the carrier proteins within a few hours. For haptens that are sparingly soluble in aqueous solutions, this reaction may be carried out in 50% DMSO or DMF. Using these succinimidyl esters, it is quite easy to prepare hapten-carriers with varying levels of hapten coupled per carrier molecule [9]. A highly substituted conjugated carrier may give rise to an antibody with strong reaction, but will be somewhat difficult to inhibit. A carrier with a single conjugated hapten might give rise to lower signals in an assay, but will be quite easy to inhibit.

The same principles hold true when conjugating these carriers to the detector particles or bind them to membranes. The developer may thus be able to tune the assay to the desired sensitivity by selecting the appropriate range of substitutions. For some haptens, substitutions at several positions of the molecule may be available. Having derivatives of these species provides a range of reagents for fine-tuning the assay to the desired sensitivity and specificity. One strategy would be to use a derivative at one position for preparing the immunogen and using other derivatives made from the substitution at this same position for immunization (homologous competitive immunoassay). Alternatively, one could use derivatives in immunoassays different from those used for immunization (heterologous competitive immunoassay). Heterologous competitive immunoassays may prove easier to inhibit by free antigen, but may not have the same high degrees of specificity as a homologous one. Specificity is a function of what closely related compounds exist and may be of concern depending on what may be encountered in the samples being analyzed. This will need to be approached on an assay-by-assay basis. In general, antibodies with very high affinity for the derivates used in the immunoassay may prove difficult to inhibit but may have the greatest degree of specificity, while antibodies with low affinity for the derivates will be much easier to inhibit but may suffer from cross-reactivity with related compounds. Careful examination of all related structures should be undertaken prior to choosing an antigen derivative for immunization and for use in the assay. Having a large repertoire of such compounds available allows for a comprehensive evaluation and maximizes the chances for success of the assay.

In preparing antigen conjugates for use in lateral flow immunoassay, it is often desirable to use a crosslinker distinct from that used for immunization [10]. In addition, the spacer arm, which is highly desirable in order to avoid steric hindrance between the antibody and the protein carrier to which the antigen is coupled, should also be different from that used for immunization. Most common crosslinking agents and spacers are immunogenic, and using the same ones for both immunization and assay may lead to complications in the assays because of mixed populations of antibodies to the conjugated antigen, linker, and spacer [11]. By the same token, the carrier proteins should also be different in the two conjugates [12]. While one could do extensive controls and cross-absorptions to attempt to remove the reactivity directed to the carrier alone, it is far easier to simply avoid the problem. Albumin, ovalbumin, or IgG make excellent carrier proteins for assay. All of these are readily used on microtiter plates for monitoring the progress of the immunizations by ELISAs. Conjugates to IgG are particularly good for binding to nitrocellulose, with albumin not binding as avidly. Both IgG and albumin couple well to the particles used as detectors, such as colloidal gold, with albumin conjugates typically being made in the pH range of 5.0–6.0 and IgG conjugates in the range of 8.0–9.0. Having both available for detector particles offers the developer some flexibility if difficulties with non-specific binding are encountered or in the pH at which the assay is conducted. Regardless of the method chosen for preparing the conjugate for assay, the conjugation procedures for linking the antigen to the carrier (as well as the coupling of this complex to the detector particles or membrane) must be highly controlled and reproducible. It should also be quantifiable. A robust small molecule conjugate will allow an assay to be developed, which does not require constant adjustment or re-tweaking.

If the protein is made by recombinant techniques, having the native material available for screening will ensure that the resultant antibodies do not have a strong bias toward the recombinant protein, which may, depending on the expression and purification system used, have a very different configuration. This phenomenon is very commonly observed. If no protein is available, but the protein sequence is known, peptides may be selected to mimic the protein [13]. Conjugation techniques as previously described will be required to produce suitable immunogen [14, 15]. While high-affinity antibodies to the immunizing peptide can be generated, they are frequently of low reactivity with native protein.

A relatively new technique for the generation of antibodies is based on the introduction of DNA coding for the antigen of interest (or portion thereof) into cells of the animals being immunized [16–19]. This is typically accomplished by bombardment of the cells at the injection site with particles coated with the DNA encoding the antigen, typically in a construct with appropriate promoters, antigen, and helper molecules to elicit a strong immune response. The DNA is transiently expressed in the cells of the animal and has the strong advantage of being mammalian in origin and expressing appropriate and proper folding. Another advantage of this technique is that no protein needs to be expressed or purified to generate the antigen. This is particularly valuable for hard-to-express proteins, proteins that are extremely labile, or proteins that are scarce in natural samples. Since no antigen purification is required, this method has the potential to make antibodies to any protein for which a sequence is known. The technique has been shown to yield extremely reproducible results. DNA immunization is equally suitable for the generation of polyclonal antibodies in mice and rabbits or for mice from which monoclonal antibodies will be generated. Antibodies from this method are transitioning their way into assays, including lateral flow. As with other methods, it is important to have the screening strategies planned in advance to ensure that an appropriately strong immune response of the targeted sensitivity and specificity has been generated. In many instances, the plasmid used for immunization can be expressed to generate protein antigen for control purposes or for affinity purification of the resultant antibodies.

4.4 Selection of the Animal Species

A commercially successful lateral flow immunoassay may require significant amounts of antibody. Consider an assay with a sales volume of 1 million units per year. If the capture (membrane immobilized) antibody is dispensed at a rate of 1 ml per ml and 1 μl per cm and the test strips are 5 mm wide (all of which are somewhat typical parameters), then over 0.5 g of capture antibody would be required per year. Selection of species and number of animals in the program needs to take this into account in the beginning. Both rabbits and goats [20] are well suited for the generation of polyclonal antibodies for lateral flow immunoassay test strips. These antibodies will likely require affinity purification using antigen in sandwich assays (described below). For competitive assays, highly purified IgG fractions may well prove adequate.

Methods for generating polyclonal antibodies involve an initial immunization, followed by a series of boosts, typically at 3–6 week intervals [21–23]. Exact dosing of antigen will vary by the route of administration, scheduling, and species of animals. Rabbits will require 0.5–2 mg per animal and goats may require 3–10 times as much for the total protocol. Antigen is usually given initially in combination with adjuvants [24, 25] such as bacterial polysaccharides (LPS), with the goal of localizing the deposit of antigen to provoke a strong immune response, as well as providing generalized stimulation of the immune response. All protocols should be reviewed by animal welfare committees to ensure they comply with the guidelines for humane treatment of the animals [26]. Exact protocols can vary, but should be designed to mimic the protocols for similar antigens or classes of antigens that have yielded successful antibodies. Companies specializing in the development of custom antibodies can generally suggest protocols for an antigen that will maximize the chances for success. Typically, 3–4 months are required to achieve antibody titers that then begin to plateau [27]. Once good antibody titers are identified, the animals can be put into production schedules consisting of a series of antigen boosts, followed several weeks later by the collection of sera. These production bleeds, on 1- to 2-month cycles, can go on for years with rabbits and goats. Animals that are producing useful antibodies should be bred continuously to collect sufficient antisera to enable a lateral flow immunoassay to stay on the market for many years, consistent with marketing forecasts for that assay. Additional animals can be introduced into the program once initial protocols have proven effective, but sera from each animal should be evaluated individually, at least by ELISA, prior to pooling or assuming adequate quality for use in lateral flow assays.

Chicken offers another excellent animal choice, which yields a repertoire of antibodies not observed with rabbits, goats, or mice [28, 29]. Antibody generated by chickens is passed from the serum into the egg yolk (IgY) and can be conveniently harvested from the eggs. Approximately 25 eggs can be obtained per chicken per month containing about 2 g of poylclonal antibodies, equivalent to 200 ml of rabbit serum. In addition, chicken IgY antibodies do not activate complement and are not bound by human rheumatoid factor, two properties that can help reduce non-specific binding in serum/plasma-based tests. Like other polyclonal antibodies, chicken antibodies will likely require affinity purification to be most useful in lateral flow tests.

Monoclonal antibodies are particularly well suited to lateral flow immunoassays. Monoclonal antibodies are the result of fusing an antibody-producing B lymphocyte, frequently from the spleen of the immunized animal, with an established myeloma cell line resulting in a "hybridoma" or fused cell that retains the property of the myeloma cell to grow indefinitely in culture and the property of the B lymphocytes to secrete antibodies [23]. Through limiting dilution cloning procedures, lines and clones can be derived from single B lymphocytes and screened for antibodies of interest by ELISA. Selected hybridomas can then be scaled up. Once a monoclonal cell line is established, many issues related to polyclonal antibodies production can be avoided. These include continuing immunizations, need for more antigen boosting, scale-up procedures, or consistency over many years. Since all the antibodies in a monoclonal preparation are directed against the antigen of interest, there is no need for antigen-based affinity purification. However, if antigen polymorphism exists, it is important to ensure that the selected hybridomas recognize the entire repertoire of targeted antigen. These antigens would likely be detected if polyclonal antibodies are used.

One of the critical factors for a successful monoclonal antibody production is the strategy used for screening the hybridomas. As cells start to grow up after limiting dilutions, usually in 96-well microplates with volumes of 200 μl or less, they will reach a point at which screening must occur to select the specific hybridoma for further expansion and sub-cloning as well as to eliminate non-specific hybridomas to reduce material and labor costs in carrying those clones. As with antigen selection, the strategy needs to be defined at the onset of the program, rather than waiting until one is needed. For this purpose, hybridomas may be selected by screening supernatants from 96-well plates based on ELISA for specificity and strength of binding to the target antigen. This will reduce the number of candidate hybridomas to be taken forward to 10–20. These hybridomas will then be expanded to obtain sufficient quantities of antibodies for evaluation through micro techniques to be described later. Several hundred micrograms would be sufficient for this evaluation.

Monoclonal antibodies are generally produced in mice and, less frequently, in rats. Methods for effective monoclonal antibody production from rabbit cells [30], where suitable myeloma lines are currently lacking as of this writing, are continuing to be refined and hopefully will soon be available.

4.5 Antibody Purifications

A variety of purification techniques are available to yield the antibodies suitable for lateral flow immunoassays [23]. The most common includes salt fractionation followed by ion-exchange chromatography, protein A (or protein G) purification, or affinity methods. There are several critical aspects that need to be considered. Firstly, the antibodies need to be very pure (>98% would be ideal). Because they are applied at relatively high concentrations, contaminant proteins may hinder maximal binding. Secondly, exposure of the antibodies to harsh conditions (extremes of pH or ionic strength) should be avoided so as to minimize denaturation.

Denaturation can result in exposure of hydrophobic residues, which in turn leads to non-specific binding on the membrane and on the detector particle. Again, this problem can be exacerbated by the high concentrations of antibodies used in lateral flow immunoassays. Lastly, the antibody preparations need to be free of any aggregates, particularly if they are to be used on detector particles. Aggregates are capable of conjugating to several particles, leading to bridging to create particle aggregates and which often create non-specific binding. It is important that aggregates are removed prior to use by centrifugation at 15,000 × g for 30 minutes. The presence of a large number of aggregates is an indication that the antibodies have experienced mild and at least mild denaturating conditions, and their use may be problematic.

Complicating the use of chicken antibodies is the high lipid content of the yolk requiring its removal prior to IgY purification. This may be accomplished using PEG precipitation or dextran sulfate. Unlike many mammalian antibodies, IgY is not bound by Protein A or protein G, which makes its purification somewhat more complicated [28].

4.5.1 Affinity Purification of Antibodies

In a typical polyclonal antibody preparation, only 0.2–2% of the total antibody are directed against the antigen of interest. The rest are not specific for the target antigen, but will compete with the specific antibodies for binding to the nitrocellulose membrane during the spraying process and for conjugation to the detector particles, effectively limiting the density of specific antibody and the sensitivity of the strip. Affinity purification can be used to effectively separate the specific antibody from the background immunoglobulins. Affinity absorbents are prepared by covalently coupling the antigen to an insoluble matrix, most commonly chemically activated agarose beads [31–34]. While these chemistries may be prepared in the laboratory, it is most convenient to purchase them as preactivated, stabilized beads. Activation chemistries include cyanogen bromide, succinimidyl esters, and epoxides for reactivities with amino and hydroxyl groups. Antigen added at pH 7.5–9.0 will couple to the particles through free amino or hydroxyl groups. After processing to block the free reactive groups on the beads, antisera or purified immunoglobulins may be added to the beads in batch or column format. It is generally advisable to use at least partially purified immunoglobulins rather than antisera to prolong the life of the absorbent. Specific antibody will bind to the antigen-coupled beads, allowing the background immunoglobulins to pass straight through. Once the non-specific immunoglobulin background has been removed (usually only a few column volumes are required), the bound antibodies may be eluted using conditions that typically disassociate the antibody–antigen complex [35, 36]. These may include, often in order of preference, low pH (2.8 or lower), high pH (11.5), or chaotropic agents, such as 3.5 M $MgCl_2$[37]. Eluted antibodies should be neutralized immediately and dialyzed into buffers of choice. Very high recovery of specific antibodies can be readily achieved. Purification factors will range from 20- to several hundred fold, depending on the fraction of total immunoglobulin that is antigen-specific. Using affinity-purified antibodies in both membrane and detector particles in sandwich assays will result in 100- to 10,000-fold increase in sensitivity over non-affinity-purified antibody. One should exercise caution when purchasing commercial antibodies labeled as "affinity purified", as this term is often misused referring to protein A-purified antibodies as opposed to the correct use of the term in referring to the antibodies purified based on its affinity for its target antigen.

To perform affinity purifications, a significant quantity of antigens, typically milligrams, will be required. For some antigens, this can be a significant obstacle. The antigen should also be stable under conditions used for elution or the immobilized antigen on beads may have a very short useful life. If purified antigen is limiting, monoclonal antibodies, which do not need affinity purification, may be the antibody of choice.

4.5.2 Cross-Absorption of Antibodies

Even with affinity purification, there may be unintended antibody specificities that co-purify with the affinity-purified antibodies. This is very common with polyclonal antibodies. The cause may be due to the epitopes shared between the antigen target and other related molecules that should not be detected by the proposed lateral flow tests. It may also result from impurities in the antigen preparation used for affinity purification. A strategy to remove the undesired activities is to use immunoabsorbents as previously described with the undesired antigen coupled to the absorbent. In this instance, the desired antibody is the fraction that does not bind to the absorbent. As with affinity purification, one should carefully monitor the reactive profile of antibody throughout the process. This will avoid over-saturating the absorbents and ensure reproducibility. ELISA is a convenient method to do this [38].

Affinity-absorption and cross-absorption are frequently combined when purifying polylconal antibodies. Whether one performs cross-absorption as the first or second step depends on the relative reactivity of the antibodies with the desired target versus the cross-reacting targets and the availability of purified target antigen and purified cross-reactants. It is usually advisable to use the material available in smaller quantities as the second step. In other words, if purified target antigen is very limiting, one should perform the cross-absorption first, followed by affinity purification. If the cross-reactant is a very minor percentage of the antibody preparation and is available in short supply, the order should be reversed.

4.6 Predictive Formatting

While techniques such as direct ELISA may provide results that allow ranking of antibodies with regard to relative affinity, it is difficult to translate this information into a general model that will predict performance in lateral flow assay. There are too many variations in the way ELISAs are configured, and the criteria imposed by the design goals of a particular lateral flow immunoassay may be much stricter than the usual ELISA criteria. Nevertheless, antibodies may be prescreened by ELISA to determine which ones will give a good sandwich pair and this may save some time in finding the appropriate ones for lateral flow immunoassays.

Dot spotting represents a method for predicting which antibodies will prove to be the most useful in sandwich lateral flow immunoassays. This technique gives quick results and is most useful when antibodies are only available in limited quantities. It works well early in a hybridoma program to help narrow down the choices of useful antibodies to proceed forward. This method would also work well for antibody samples from commercial vendors. All antibodies to be evaluated need to be purified and should be at concentrations of 0.5 mg/ml or more. This is well within the realm of what could be expected from 10 to 20 ml of culture supernatants from the hybridomas being expanded in a monoclonal program. The antibodies can be readily purified with a 1-ml Protein A Sepharose column using standard methods [23], followed by dialysis in phosphate-buffered saline, pH 7.4. Concentration of the resultant antibodies, though not usually required, can be accomplished using a centrifugal concentrator.

Conditions for optimizing antibody conjugation to colloidal gold [39–41] (also see Chapter 5) may be conveniently evaluated in microtiter plate wells. For this purpose, 100 μl of colloidal gold at 1 OD_{530} are added to each well followed by 10 μl of 22 mM buffers at pH from 5.5 to 10 in 0.5 pH unit increments, resulting in a final concentration of 2 mM buffer within the well. Suitable buffers are MES from pH 5.5 to 6.5, HEPES from 7.0 to 7.5, and Borate from 8.0 to 10. 10 μl of antibodies, diluted in 2 mM Borate pH 9.0 at 100, 50, 37.5, 25, 18.75, and 12.5 μg per ml is added with immediate gentle mixing upon addition. This will yield final concentrations of 10, 7.5, 5.0, 3.75, 2.5, 1.875, and 1.25 μg of antibodies per 1 OD of colloid, respectively. This mixture is allowed to stand for 15 min, followed by the addition of 25 μl of 1.5 M NaCl. Colloidal particles conjugated with sufficient antibody will be stable and remain pink, whereas colloid with insufficient stabilization will aggregate and turn pale blue. Optimal conjugation condition is the pH range where the lowest concentration of antibodies provides protection against aggregation. These conditions can be readily scaled up to assay volumes of 1 or 2 ml. By selecting the pH conditions with the minimal amount of antibodies required for stabilization, one may be able to eliminate the washing step to remove free antibodies, thus making scaling up easier. Bovine serum albumin and polyethyleneglycol (20,000 molecular weight) should be added to a final concentration of 0.1% to further stabilize the colloid when working with compounds of lower molecular weight than the immunoglobulins. This material may be used directly for dot spot analysis. Conditions used in this optimization usually scale up very well. For large-scale preparations, one should take additional steps to ensure that no free antibodies remain.

Typically, in lateral flow immunoassays, lines are applied to membranes using dispensing equipment, which, depending on the equipment, frequently requires significant amounts of antibodies for priming. For a quick analysis of many antibodies, they can be applied to the membrane in spots with a pipette capable of dispensing 1 μl. These antibodies should be at a concentration of 0.1 mg per ml or greater. The novice spotter may find that adding 0.1–1% food coloring to track the zones of antibodies will help. The food coloring does not interfere with antibody binding and will wash out during testing. Membranes should be allowed to dry for a minimum of 1 hr at 37°. Mocked-up lateral flow test strips may be made using a vinyl backing, the spotted nitrocellulose, and an absorbent pad. No conjugate pad or sample pad is used. To perform this test, 50–100 μl of test antigen or buffer control is mixed with 10–50 μl of the colloidal gold antibody preparation of another antibody aimed at finding a potential sandwich pair. It is generally advantageous to have surfactant and some proteins in the buffer solution (e.g., 0.1% Tween 20 and 0.1–1% bovine serum albumin). The strips are placed directly into the antigen gold solution and allowed to develop for 5–10 minutes. All antibodies in this evaluation should be tested as both the capture antibodies (membrane antibodies) and all the antibodies for the detector particles. While optimized pairs will frequently work in both configurations, some tests may work better with a preferred orientation. Non-specific interactions will appear as completely filled-in dots in the presence of buffer alone. Very weak non-specific interactions can often be eliminated later in the development process through adjustment of antibodies, conjugates, and buffers. Occasionally, strong non-specific interactions may be observed between certain antibody pairs. These interactions may be so strong as to occur only at the leading edge of the applied antibody dot. These reactions are attributable to charge interaction on the antibodies. In practice, such pairs are best avoided as attempts to satisfactorily prevent the non-specific interactions usually prove difficult, if not impossible. Ideally, one should select pairs that have no hint of non-specific interactions.

When antigen is tested in the dot spotting system, one would see dots filled in with conjugates. Strong high-affinity interactions, which are often the most desirable, would only show a test reaction at the leading edge of the dot, indicating that all the antigens are being effectively captured as soon as they enter the test zone. Antigen titrations can be easily done in this format and the useful antibody pairs can readily be identified for expansion, scale up, or further purchase.

The same principles can be applied to evaluating antibodies for competitive assays, although both antibody and antigen concentrations (regardless of which is used for capture or detection) have to be titrated. The goal is very similar to establishing an ELISA competitive assay to find the dilution of each reagent such that a change in concentration of that reagent produces a change in the signal of the assay in the absence of free antigen. This would then be the starting point for attempting inhibitions with free antigens.

4.7 Quality Control

Since variations are inherent in all biological systems, assuring an uninterrupted supply of consistent antibodies is tantamount in maintaining the long-term ability of the developer to continuously produce lateral flow devices. Intuitively, one would think that monoclonal antibodies are the most consistent, being produced from a single cloned cell line. However, variations can exist here too, most likely as a result of the production and purification methods employed. Production of monoclonal antibodies by in vitro methods versus ascites can influence the glycosylation of the antibodies that, with altered charges, may affect their binding properties to nitrocellulose or to the detector particles. In this respect, use of isoelectric focusing (IEF) as a quality control step provides safeguards against such undesirable changes. Most monoclonal antibodies exhibit a narrow range of several bands detectable by IEF, and the key parameter to monitor is whether new bands appear or existing ones get deleted. Bands appearing significantly outside of the defined range may require re-optimization to match performance. Additionally, SDS polyacrylamide gel electrophoresis should be used to guarantee a purity of greater than 98%. Polyclonal antibodies do not lend themselves well to isoelectric focusing as a quality control step. Nevertheless, SDS polyacrylamide electrophoresis should be done to confirm purity.

With polyclonal antibodies, particularly affinity-purified ones, maintaining consistent specific activity is a critical parameter. If the affinity columns used for production change over time, a drift may occur in the specific activity. These changes may arise as a result of repeated cycles of the mildly denaturing conditions used for elution. Drift is also often observed when new columns are generated, particularly if they involve new antigen preparations. Additionally, there may be drift with regard to cross-reactivity if the columns used for cross-absorption deteriorate. ELISA provides a reasonable, though not always foolproof, method for assessing specific activity. A significant decrease in specific activity is often indicative of a change in the affinity column. As a result, more antibodies would be required in the final lateral flow tests, which may not always be feasible. Increases in specific activity are not always desirable either, since this would require some re-optimization of the conditions to maintain the desired consistent sensitivity. A cross-reactive panel should also be run by ELISA, using the same antigens or organisms originally screened for.

The above precautions should minimize the introduction of variation in the production of antibodies for lateral flow immunoassays. However, as a final safeguard, it is always useful to test the newly prepared reagents against a previous lot of antibodies to guarantee identical performance. While it is obvious that one should sequester amounts of previous lots for this quality control step, it is amazing how often this step is overlooked. In addition, if new animals are immunized and brought into the polyclonal program, their antisera and purified antibodies will need to be carefully qualified before assuming that they are suitable replacements for the original ones. If all animals in the original immunization program respond well and produce useful antisera, it is reasonable to assume good chances of success with subsequent animals, if identical antigen and immunization regimens are used. If, however, only a small proportion of the original animals produce the desired response or if the antigen is from a different lot, the system has to be re-characterized every time new animals are introduced. In situations where the animals respond with the desired antibody titers, it is advisable to implement a long-term production strategy, consisting of alternating boosts and bleeds and to stockpile the raw antisera to guarantee several years' worth of antibodies. Sera may be stored frozen at −80°C for decades, and can be thawed and purified as needed.

4.8 Conclusion

There are many components that make up a lateral flow immunoassay test. The antibody component is probably the most important. It is a biological product that is inherently variable. Carefully designed pre-screening and purification programs should be in place before the start of an antibody production project. Strategies should also be in place to assure continuous and consistent supply of antibodies with proven affinity and specificity. One approach is the use of monoclonal antibodies, which allows production of specific antibodies in large quantities. Currently, most monoclonal antibodies are derived from mouse hybridomas. Rabbit cell hybridomas are not yet available. As with other production strategies, quality control of antibody preparations is an important aspect of constructing a robust lateral flow immunoassay.

References

1. Rapid Lateral Flow Test Strips, Considerations for Product Development. Bedford, MA: Millipore Corporation, (2001).
2. Nunc Guide to Solid Phase, Rowell, V. Ed., Roskilde, Denmark: Nunc A/S, Jan. 2001.

3. Chandler, J., Gurmin, T., Robinson, N. (2000) The place of gold in rapid tests. IVD Technology 6(2):37–49.
4. Linscott, D.W. Linscott's Directory of Immunological and Biological Reagents. (2008) http://www.linscottsdirectory.com
5. Layton, G.T., Stanworth, D.R. and Amos, H.E. (1986) Factors influencing the immunogenicity of the haptenic drug chlorhexidine in mice. II. The role of the carrier and adjuvants in the induction of IgE and IgG anti-hapten responses. Immunology 59(3):459–465.
6. Harris, J.R. and Mark, J. (1999) Keyhole limpet hemocyanin (KLH): a biomedical review. Micron 30(6):597–623.
7. Wong, S.S. (1991) Chemistry of Protein Conjugation and Cross-Linking. Florida: CRC Press.
8. Cuatrecasas, P. and Parikh, I. (1972) Adsorbants for affinity chromatography. Use of N-hydroxysuccinimidyl esters of agarose. J. Biol. Chem. 11(12):2291–2299.
9. Pedersen, M.K., Sorensen, N.S., Heegaard, P.M., Beyer, N.H. and Bruun, L. (2006) Effect of different hapten-carrier conjugation ratios and molecular orientations on antibody affinity against a peptide antigen. J. Immunol. Methods 311(1–2):198–206.
10. Carter, J.M. (1994) Techniques for conjugation of synthetic peptides to carrier molecules. Methods Mol. Biol. 36:155–191.
11. Gullick, W.J. (1994) Production of antisera to synthetic peptides. Methods Mol. Biol. 32:389–399.
12. Walter, G. (1986) Production and use of antibodies against synthetic peptides. J. Immunol. Methods 88(2):149–161.
13. Hancock, D.C. and Evan, G.I. (1998) Production and characterization of antibodies against synthetic peptides. Methods Mol. Biol. 80:15–22.
14. Meloen, R.H., Puijk, W.C., Langeveld, J.P., Langedijk, J.P. and Timmerman, P. (2003) Design of synthetic peptides for diagnostics. Curr. Protein Pept. Sci. 4(4):253–260.
15. Adrian, T.E. (1997) Production of antisera using peptide conjugates. Methods Mol. Biol. 73:239–249.
16. Chambers, R.S. and Johnston, S.A. (2003) High-level generation of polyclonal antibodies by genetic immunization. Nat. Biotechnol. 21(9):1088–1092.
17. Tang, D.C., DeVit, M. and Johnston, S.A. (1992) Genetic immunization is a simple method for eliciting an immune response. Nature 356:152–154.
18. Thalhamer, J., Leitner, W., Hammer, P. and Brtko, J. (2001) Designing immune responses with genetic immunization and immunostimulatory DNA sequences. Endocr. Regul. 35(3):143–166.
19. Cohen, A.D., Boyer, J.D. and Weiner, D.B. (1998) Modulating the immune response to genetic immunization. FASEB J. 12(15):1611–1626.
20. Bailey, G.S. (1994) The raising of a polyclonal antiserum to a protein. Methods Mol. Biol. 32:381–388.
21. Hurn, B.A. and Chantler, S.M. (1980) Production of reagent antibodies. Methods Enzymol. 70(A):104–142.
22. Drenckhahn, D., Jons, T. and Schmitz, F. (1993) Production of polyclonal antibodies against proteins and peptides. Methods Cell Biol. 37:7–56.
23. Harlow, E. and Lane, D. (1988) Using Antibodies: A Laboratory Manual. Cold Spring Harbor Laboratory, Cold Spring Harbor, New York.
24. Johnson, A.G. (1994) Molecular adjuvants and immunomodulators: new approaches to immunization. Clin Microbiol. Rev. 7(3):277–289.
25. Stills, H.F., Jr. (2005) Adjuvants and antibody production: dispelling the myths associated with Freund's complete and other adjuvants. ILAR J. 46(3):280–293.
26. Leenaars, M. and Hendriksen, C.F. (2005) Critical steps in the production of polyclonal and monoclonal antibodies: evaluation and recommendations. ILAR J. 46(3):269–279.
27. Schunk, M.K. and Macallum, G.E. (2005) Applications and optimization of immunization procedures. ILAR J. 46(3):241–257.

28. Tini, M., Jewell, U.R., Camenisch, G., Chilov, D. and Gassmann, M. (2002) Generation and application of chicken egg-yolk antibodies. Comp. Biochem. Physiol. A Mol. Integr. Physiol. 131(3):569–574.
29. Carlander, D., Stalberg, J. and Larsson, A. (1999) Chicken antibodies: a clinical chemistry perspective. Ups. J. Med. Sci. 104(3):179–189.
30. Saito, M., Sakurai, S., Motegi, A., Saito, K., Sano, T. and Nakajima, T. (2007) Comparative study using rabbit-derived polyclonal, mouse-derived monoclonal, and rabbit-derived monoclonal antibodies for KIT immunostaining in GIST and other tumors. Pathol. Int. 57(4):200–204.
31. Fisher, P.A. and Smith, D.E. (1988) Affinity purification of antibodies using antigens immobilized on solid supports. Biochem. Soc. Trans. 16(2):134–138.
32. Hermanson, G.T., Mallia, A.K. and Smith, P.K. (1992) Immobilized Affinity Ligand Technique. California: Academic Press.
33. Dean, P.D.G., Johnson, W. S. and Middle, F.A. (1985) Affinity Chromatography: A Practical Approach. Oxford: IRI Press.
34. Jack, G.W. (1994) related articles, immunoaffinity chromatography. Mol. Biotechnol. 1(1):59–86.
35. Fornstedt, N. (1984) Affinity chromatographic studies on antigen-antibody dissociation. FEBS Lett. 177(2):195–199.
36. Hoffken, K., Bosse, F., Steih, U. and Schmidt, C.G. (1982) Dissociation and isolation of antigen and antibody from immune complexes. J. Immunol. Methods 53(1):51–59.
37. Tsang, V.C. and Wilkins, P.P. (1991) Optimum dissociating condition for immunoaffinity and preferential isolation of antibodies with high specific activity. J. Immunol. Methods 138(2):291–299.
38. Kummer, A. and Li-Chan, E.C. (1998) Application of an ELISA-elution assay as a screening tool for dissociation of yolk antibody-antigen complexes. J. Immunol. Methods 211(1–2):125–137.
39. De Roe, C., Courtroy, P.J. and Baudhuin, P. (1987) A model of protein-colloidal gold interactions. J. Histochem. Cytochem. 35(11):1191–1198.
40. Horisberger, M., Rosset, J. and Bauer, H. (1975) Colloidal gold granules as markers for cell surface receptors in the scanning electron microscope. Experimentia 31:1147–1149.
41. De May, J. (1986) Colloidal gold probes. In: Pollack, J. and van Noorden, S., Eds. Immunochemistry, Modern Methods and Applications, Bristol: Wright PSG press, pp. 82–92.

Chapter 5
Colloidal Gold and Other Labels for Lateral Flow Immunoassays

Peter Chun

5.1 Introduction

Lateral flow immunoassays have seen widespread applications, as witnessed by a compendium of reviews in the literature. In this chapter, the author describes some of the labels that have been successfully commercialized and surveys others that appear to be promising. The discussions are equally applicable to nucleic acid lateral flow immunoassays. More emphasis has been put on the sections on colloidal gold because of its popularity and because of the author's background. Compared to other chapters in this book, the current chapter describes more detailed procedures. Hopefully, the reader can utilize some of the protocols without having to resort to the original literature.

The development of labels or markers for lateral flow immunoassays has gone hand in hand with the advances in detection methodology and instrumentation. Very sensitive assays using fluorescent and luminescent labels have appeared in recent years, initially as non-membrane-based heterogeneous or homogeneous immunoassays. Over time, these have been adapted into lateral flow formats. For these labels and their corresponding detectors, it is essential to remember several performance requirements including wavelength, sensitivity, dynamic range, stability, time resolution, and costs.

The ideal label for lateral flow tests would possess the following characteristics:

a. Amenable to detection by multiple methods or technologies over a very large and useful dynamic range.
b. Simple conjugation chemistries are available so that biologicals and chemicals can be conjugated without loss of chemical and biological integrity and activity.
c. Possesses no or low non-specific binding characteristic such that a high signal-to-noise ratio under many salt, buffer, and detergent conditions.
d. Stable under various chemical conditions and temperatures.

P. Chun (✉)
EASE Laboratories Inc., South San Francisco, CA
e-mail: pechun2@yahoo.com

R.C. Wong, H.Y. Tse (eds.), *Lateral Flow Immunoassay*,
DOI 10.1007/978-1-59745-240-3_5, © Humana Press, New York, NY 2009

e. Commercially available at low cost.
f. Conjugation procedure is easy, cost-effective, and scalable.
g. Easy to process downstream.
h. Capable of being used for multi-analyte detection.
i. Usable for anticipated advancements in technology.

However, a label that possesses all these characteristics does not exist. One then has to choose a label based on the following conditions of use:

i) The detection/visualization method.
ii) The analyte to be detected.
iii) The detection range.
iv) The environment under which the test has to be performed, and stored.
v) The biologics or chemicals that must be bound onto the label, and the available chemistry that can be used to avoid alteration to its conformation or steric hindrance.
vi) Quantity and quality of materials available.
vii) Cost considerations.
viii) Future needs.

Based on the above considerations, descriptions on some of the labels are presented in the following sections.

5.2 Liposomes

Liposomes are vesicles formed by a lipid bilayer and have been used commercially in membrane-based assays in vertical and lateral flow formats. (e.g., test for malarial antigen from Becton Dickinson). Because of the ability to encapsulate very high concentrations of signal generating molecules within their cores, the diagnostic application of liposomes can improve a visual lateral flow immunoassay's test sensitivity by 2–3 orders of magnitude.

Depending on the detection method used and the sensitivity of the test required, liposomes, ranging in sizes from 50 nm to 800 nm, can be made to encapsulate visual dyes, fluorescent dyes, enzymes, or electroactive compounds. Lipoproteins, glycolipids, and various other lipid-containing compounds can be incorporated directly into the bilayer. In addition, different chemically active groups can be incorporated on the lipid surface with controlled surface density for covalent coupling to biological or chemicals [1].

Liposomes can be lysed to release the encapsulated material within leading to detection. For example, with an encapsulated electroactive compound, lysis of the liposome with appropriate surfactants results in the release of the encapsulant that, when comes into contact with an interdigitated ultamicroelectrode array (IDUA) microchip applied with a small electrical potential, would generate a detectable electric current.

However, one of the drawbacks of liposomes is its relative instability and susceptibility to lysis by surfactants although a commercial liposome manufacturer claims that 36 months of stability is possible. Another disadvantage is that there is scant literature available for the drying and reconstitution of liposomes, thus making it difficult to manufacture true "one-step" self-contained tests. Because of these difficulties, most of the commercially successful tests employing liposomes have been liquid reagents and not lateral flow immunoassays.

5.3 Colloidal Carbon

The use of colloidal carbon in the form of "India Ink" in immunoassays has been reported since the 1970s [2–4]. However, due to the inconsistency of India Ink, it was difficult to produce commercial immunoassays. In 1993, van Amerongen et al. [5] described the use of colloidal carbon particles as a new label for rapid immunochemical test methods, and discussed its possible use in lateral flow immunoassays. In 1995, a patent by Jacob Van Doorn of Staatder Nederlande [6] describes how unstabilized carbon could be used to produce carbon sols suitable for protein adsorption. Their carbon sols were formed by suspending carbon particles of well-defined particle sizes in distilled water or low ionic strength buffers, sonicated or vigorously agitated, followed by centrifugation. These unstabilized carbon sols were flocculated easily by salt. However, when coated with macromolecules such as antibodies, they were "protected" from flocculation. In practice, increasing amounts of a macromolecule are incubated with a fixed amount of non-stabilized carbon aqueous sol under defined conditions to determine the "minimal protective amount". The optimal pH for adsorption is determined empirically. Unlike colloidal gold, in which the conjugation of protein to colloidal gold is near instantaneous, adsorption onto colloidal carbon takes a longer time from one to several hours.

Some of the advantages of colloidal carbon include its good stability and high color contrast on a membrane. It is fairly easy to conjugate and economical such that a bottle of carbon particles lasts for millions of tests. Currently, there are only a few vendors for well-characterized carbon particles and the use of colloidal carbon for lateral flow assays requires a licensing agreement from the vendors making it less attractive than colloidal gold to be used commercially. As with colloidal gold, multiplexing is only achievable with spatial differentiation on the membrane. Several hormone and infectious disease lateral flow tests have appeared on the market using colloidal carbon as a label. These include "Best Test" and "Contrast" series of rapid tests from Genzyme Diagnostics (San Carlos, CA 94070) and marketed by Becton Dickinson.

5.4 Colloidal Gold

Colloidal gold is perhaps the most widely used label today in commercial lateral flow immunoassays for many reasons [7]. It is fairly easy and inexpensive to prepare in the laboratory. The color is intense, and no development process is need for visualization. A large body of protocols exists in the literature for its conjugation and applications. The label is very stable in liquid or dried form and is non-bleaching after staining on membranes. In addition, colloidal gold in unconjugated forms (which are ready for labeling) and conjugated forms (conjugated with biologicals) are now readily available from many commercial sources such that a quick search on the Internet can easily yield more than 20 vendors.

Broadly defined, colloids are a stable dispersion of one phase in another. The term colloidal, derived from the Greek word for glue, was coined in 1861 by the chemist Thomas Graham. About this time, Enrico Selmi precisely described the colloid phenomenon and developed a theory of charge interaction of which several aspects are still valid today. One of the first scientific studies on the preparation and properties of colloidal gold was done by Faraday in 1857. He discovered the sensitivity of colloidal gold to electrolytes and demonstrated the protective effect of gelatin and other macromolecules. In the early twentieth century, colloidal gold re-emerged within the field of particle science as an investigational solution used to examine the aspects of colloidal charge repulsion, nucleation, and condensation. Major advances were made, particularly in the methods used for synthesis of colloidal gold sols. The composition of colloidal gold particles was determined to consist of an elemental gold core surrounded by a negative, ionic double layer of charges.

In 1971, Faulk and Taylor [8] first introduced the colloidal gold produced by reduction using white phosphorus for immunoelectron microscopy. However, it was not until Frens [9] described a simple sodium citrate reduction method for producing colloidal gold solution of uniform and controllable size, that sets the stage for the widespread use of colloidal gold as a label in the biological field. The first use of an antibody (see Chapter 4) conjugate colloidal gold reagent for a diagnostic immunoassay was reported by Leuvering et al. in 1981 [10]. In 1984, Brada and Roth [11] and Moeremans et al. [12] reported the use of Immunogold and Protein A-gold conjugates in Western blots and dot blots on nitrocellulose membranes. Today, it is one of the most commonly used labels for lateral flow immunoassays.

5.4.1 Preparation of Colloidal Gold

There are various ways for the production of colloidal gold in the laboratory depending on the particle size desired. An excellent review is available from Handley [13]. Basically, all methods use a reducing agent to convert ionic gold into metallic gold in a controlled manner. The reducing agents used include sodium borohydride, white phosphorus, ethyl alcohol, ascorbic acid, sodium citrate, and citrate plus tannic acid. Among them, the most commonly used method is the sodium citrate procedure developed by Frens. His procedure is briefly described below.

To a boiling or near boiling solution of tetrachloroauric acid is added a solution of sodium citrate. Immediately, gold atoms start to form in the solution and their concentration rises rapidly until the solution reaches supersaturation. Aggregation subsequently occurs in a processed called nucleation, with central icosahedral gold cores of 11 atoms forming at nucleation sites. This nucleation process occurs very quickly. Once it is achieved, the remaining dissolved gold atoms continue to bind to the nucleation sites until all atoms are removed from solution. The number of nuclei formed initially determines how many particles finally grow in solution, and the more the nuclei the smaller the gold particles would be produced. Goodman et al. has systematically graphed the ratio of particle diameter to citrate [14].

Some practical aspects of colloidal gold synthesis are presented below:

i) Glassware used should be very clean. Avoid scratching the inner surface of the glassware during cleaning by using a mild detergent and a soft cloth or tissue to scrub the inner surface. The glassware needs to be thoroughly rinsed with distilled water to remove all traces of detergents. In the author's laboratory, the rinsing process is followed by boiling with distilled water two times.
ii) Some researchers have advocated siliconizing the inner surface of the glassware to be used for preparing colloidal gold. The author had used both treated and untreated glassware extensively, and the results have basically been the same. However, siliconized glassware has a much lower tendency to be coated with unconjugated gold. Hence, it is generally easier to clean.
iii) All solutions and the distilled water used should be filtered to remove particulates or lint, which will adversely affect the uniformity of colloidal gold during production.
iv) It is best to add sodium citrate solution to a boiling or near boiling solution of gold chloride than vice versa. The sodium citrate should be added quickly and all at one time into the gold chloride solution in strong agitation. The agitation can be achieved with a vigorous mechanical agitator. For smaller volume of up to 2 L, a magnetic stirrer can be used.

In the author's laboratory, typically 1 L of distilled water is put into a clean 2 L Erlenmeyer flask, which has been pre-weighed and brought to boil using a hot plate. A glass beaker is put over the opening of the flask to minimize loss due to evaporation. Then, 10 ml of a 1% gold chloride solution is added to the water and agitated by swirling manually. A light yellow color can be seen. The solution is returned to the hot plate for continued heating at lower temperature. A pre-measured amount of 2% sodium citrate (the volume varies depending on the particle size desired) is then added rapidly all at one time into the boiling solution of gold chloride that is being vigorously agitated manually. The container is then returned to the hot plate. The solution turns initially from gray to dark gray color, then to a purple, and finally a red color. After heating for another 10 min to ensure completion, the solution is allowed to cool to room temperature. The final amount of colloidal gold sol is determined by subtracting the initial weight of the Erlenmeyer from the total weight of the flask plus liquid at the end of the reaction.

The relative concentration of each batch of colloidal gold can be determined by absorbance at 520 nm. Various batches can be brought to the same relative concentration by the addition of de-ionized water. The author has found it easier to produce multiple smaller batches made under identical conditions and procedure and combine them rather than to produce a single large batch at one time.

All gold sols display a single absorption peak in the visible range between 510 and 550 nm, and the absorption maximum shifts to a longer wavelength with increasing particle size. The relative uniformity of the particles or the range of particles can be gauged by the width of the absorption spectra: the sharper the band, the more uniform the particles.

5.4.2 Preparation of Colloidal Gold Conjugates

The composition of colloidal gold particles was determined to consist of an elemental gold core surrounded by a negative, ionic double layer of charges. In the native colloidal solution, high concentration of electrolytes causes particle flocculation and loss of the characteristic red color to a blue gray color. This is due to the electrostatic repulsive charges between particles being reduced by the electrolyte.

Macromolecular ligands adsorb onto colloidal gold through a combination of electrostatic and hydrophobic interactions [15, 16]. Colloidal gold with a suitable amount of adsorbed macromolecules are not subject to flocculation with high salt, making this an excellent tool for determining whether the colloid has been sufficiently "protected".

During colloidal gold conjugation, it is important to control the pH of the ligand and colloidal gold. Both preparations should be adjusted to a pH slightly above the isoelectric point of the ligand before conjugation. Below the pKi of the ligand, ligand-induced flocculation will occur, whereas above the pKi of the ligand, there is limited adsorption due to charge repulsion between the ligand and the colloid. The pH dependence applies primarily to protein ligands, and not to polyglycols. This property has led to the use of polyethylene glycols to stabilize colloidal gold.

The optimal pH value for conjugation to the gold sol can be determined by a method similar to that reported by Horisberger et al. [17]. Series of 0.5 ml of colloidal gold suspensions having different pH values (increasing by 0.5 pH units) are added to a series of aqueous solutions (50 μl) of protein at about 1 mg/ml. Flocculation of the gold sol will occur in some of the preparations as witnessed by a shift from red to gray purple color. The smallest pH value at which flocculation does not occur will be the optimal pH for the stabilization of the gold sol.

The minimal amount of protein required for the stabilization of the colloidal gold can be determined by adding a constant volume (e.g., 100 μl) of colloidal gold of a determined size at the optimal pH of conjugation to a 100 μl of serial dilutions of the protein also at the optimal pH. After mixing, 100 μl of 10% sodium chloride is added and the formation of flocculation is determined visually. The minimum amount of protein that is able to prevent flocculation is considered as the minimum amount of protein required to stabilize the gold sol. The amount of protein necessary to protect a larger amount of gold sol can thus be calculated. An important point to consider is that the minimum amount of protein required to stabilize the gold sol does not represent saturation values. A 10–20% or even a few folds excess can be used to attain maximum binding. In this case, excess ligand must be removed, otherwise it would compete with the gold-conjugated ligand during an assay.

The author has found that a pH of 7.5 is generally acceptable for many monoclonal antibodies (see Chapter 4). HEPES buffer is very convenient to use, since it dissolves to give a pH of 7.5 without the need to titrate the solution further. Antibodies dialyzed against or buffer exchanged on a column to 10 mM HEPES works well. Gold sols adjusted with HEPES are also stable. To adjust the pH of gold sols, a gel-filled pH meter electrode is preferred since unconjugated colloidal gold will bind to and foul regular glass electrodes. If a standard electrode has to be used, a small amount (between 50 and 100 μl) of polyethylene glycol (molecular weight 20,000 dalton) can be added to the unconjugated gold before reading the pH value. The amount of buffer required to reach the desired pH is noted. It is important to realize that the aliquot containing PEG cannot be used for conjugation and should be discarded. A proportional amount of buffer is then added to the unstabilized stock preparation to achieve the proper pH before the conjugation process. Potassium carbonate, potassium chloride, and dilute hydrochloric acid can also be used to titrate colloidal gold.

The use of very concentrated proteins during the conjugation process should be avoided as this may lead to flocculation. In the author's laboratory, the protein is dissolved in buffer to about one tenth the volume of colloidal gold to be used and filtered through a 0.2 μ filter before conjugation to remove particulate matter, since they may give rise to unsatisfactory and unstable conjugates. The preferred order of addition of the two reagents is gold sol to protein solution to prevent the formation of aggregates.

Various types of concoctions, some proprietary, have been added to stabilize the gold conjugates. Generally, they contain polyols and proteins. The use of any additives containing sulfhydryls should be avoided since they destabilize the binding of protein to colloidal gold, as they may disrupt the bonding of cysteine residues in proteins to gold. Serum albumins, such as bovine serum albumin, have been found to be good stabilizers. However, it is important to use protease-free bovine serum albumin preparations. Sodium azide up to 0.1% can be used as a preservative.

Inevitably, the colloidal gold conjugate needs to be processed further to remove excess ligand and to adjust its concentration. This is mostly accomplished by centrifugation. In general, a low-speed centrifugation is used to remove large particles before high-speed centrifugation is applied. Centrifugation protocols reported in the literature, especially those with reference to *g* values, should only be taken as guidelines. The actual results depend very heavily on the rotor geometry. One should avoid harsh centrifugation as the particles may become too packed, although in general a well-prepared conjugate can withstand a good amount of centrifugation and still be resuspended well. Pellets or parts of pellets that do not resuspend easily should be discarded since they are invariably aggregates that will give a lot of problems if used in immunoassays.

An alternative method used for concentrating colloidal gold conjugate is the use of tangential flow membranes. When using this method, it is preferable to first pre-wet the membranes with a solution containing bovine serum albumin to prevent excessive loss of the conjugate.

To minimize the amount of solution to be centrifuged after conjugation, concentrated native unconjugated colloidal gold can be used as the starting material. These are commercially available from several colloidal gold vendors, or can be prepared by concentrating the unconjugated gold before use. Concentrating the colloidal gold before conjugation has the advantage of smaller and more wieldy volumes during the conjugation process and the use of less buffers and stabilizers. However, one should be careful not to use too highly concentrated gold, as the reproducibility seems to be difficult to control and more aggregation can occur.

The processing of colloidal gold conjugates for lateral flow immunoassays, including its impregnation and drying, are beyond the scope of this discussion (see Chapter 8). Excellent reviews and protocols can be found in the literature, including manuals and product descriptions from commercial vendors of lateral flow accessories (such as conjugate pad manufacturers) and manufacturers of dispensers and automated manufacturing equipment [18–20].

5.4.3 Silver Enhancement of Colloidal Gold

A discussiosn of colloidal gold would not be complete without mention of silver enhancement. Danscher [21, 22] and Holgate [23] first described practical methods using autometallography for the enhancement of immuno-colloidal gold stained tissue sections. Silver lactate or acetate is used as the ion source and hydroquinone dissolved in a citrate buffer of low pH is used as the reducing agent. The general process results in the creation of shells of metallic silver around colloidal gold particles, leading to enhanced visibility of the colloidal gold. These preparations have to be shielded from bright daylight during the stain development procedure. After the desired level of enhancement is achieved, fixing solutions are usually used to stabilize the results.

A typical protocol (adapted from Hacker et al. [24]) at the author's laboratory for utilizing the silver enhancement in lateral flow immunoassay is as follows.

1. Silver acetate solution (Solution #1) is made fresh by dissolving 100 mg of silver acetate in 50 ml of distilled water.
2. Citrate buffer is made by dissolving 23.5 g of trisodium citrate dihydrate and 25.5 g citric acid monohydrate into 850 ml of distilled water.
3. 250 mg hydroquinone is dissolved in 50 ml of citrate buffer and freshly adjusted to pH 3.8 using citric acid. This is solution #2.
4. Just before use, an equal volume of solution #1 and #2 are mixed to form solution #3.
5. Strips of nitrocellulose are thoroughly washed with distilled water.
6. The nitrocellulose strip is dipped into solution #3 for 2–5 min.
7. As soon as the desired level of intensification is reached, further staining is stopped by dipping the strip into a readily available photographic fixer diluted 1:10 (e.g., Amphix, Ilfospeed, or Agefix).
8. The strip is then rinsed with distilled water to remove the chemicals.

It has been reported that the sensitivity of conventional colloidal gold lateral flow immunoassays can be enhanced by one to two orders of magnitude by using a silver enhancement step [25]. In their study, Horton and coworkers washed the completed lateral flow immunoassay in phosphate-buffered saline and Tween-20 solution, and immersed it in a silver enhancer reagent for 5 min. Sensitivity was reported to increase from 100 ng/ml before the enhancement to 100 pg/ml after the enhancement. Typical sensitivities for antigen detection using lateral flow immunoassay today are in the low nanogram range, and a silver enhancement would increase this to the 10 pg range of detection.

5.5 Fluorescent Probes

Fluorescence is the physical phenomenon that occurs when a molecule absorbs light at one wavelength and re-emits light at a longer wavelength. Absorption of a quantum of light energy by a molecule causes the transition of an electron from the singlet ground state to one of a number of possible vibrational levels of its first singlet state. There are several ways this electron can return to its original state, including: (a) fluorescence; (b) thermal vibrational deactivation; (c) dynamic quenching; (d) phosphorescence; and (e) resonance energy transfer. The average interval between absorption and emission is the fluorescence lifetime of a fluorescent molecule (fluorophore). Most conventional fluorophores have lifetimes of 1–10 nanosecond. A few molecules have much longer lifetimes with up to 2 microseconds for some lanthanide complexes. Lifetimes are very useful because they can be used to obtain information about the molecular motion taking place on the same time scale, and the availability of fluorophores with long lifetimes permits the assays based on time-resolved methods to be developed. Another important term for fluorophore is its quantum yield or efficiency, which is defined as the fraction (or %) of photons absorbed by a fluorophore that are re-emitted as fluorescence. A compendium of useful fluorescent labels and their characteristics can be found in various commercial catalogs, which are also excellent resources for procedures [26]. The decision on which fluorophore to use depends on the following properties: excitation and emission wavelengths, stokes shift, lifetime, extinction coefficient, quantum yield, photostability, and physical–chemical properties, including solubility and linking chemistries.

The amount of fluorescence produced is dependent upon the intensity of the excitation source. With the use of lasers, fluorescent labels can be detected at very low concentrations. Often, this is offset by photobleaching, which is exacerbated by dissolved oxygen and the presence of redox-active species.

The xanthene dyes (such as fluorescein and Texas Red) and the long-wavelength Cyanine (Cy) dyes are some of the most widely used fluorophores in general biology as well as in lateral flow immunoassays. These dyes are available commercially (e.g., I-Chroma in Soeul, Korea and Response Biomedical Corp. in Canada) with the reactive groups enabling easy conjugation to the amino group of proteins (e.g., isothiocyanate and NHS esters) [27].

In order to create highly fluorescent tags, high fluorophore:protein ratios are frequently employed during conjugation. This approach is useful only to a certain extent because extensive modification of the protein frequently cause physicochemical changes that can lead to changes in hydrophobicity, conformational changes, and steric hindrance. Ultimately, binding of the conjugate is affected. Another problem is a phenomenon called concentration quenching. This can occur when fluorescent labels are in such close proximity that when excited, less radiation energy transfer occurs and thus lowering the quantum yield. These problems can be alleviated to a large extent by first conjugating high densities of fluorophores to carrier molecules, and then conjugating the antibodies to this fluorescent complex. Compared with direct conjugation of fluorescent dyes to antibodies, the molar ratio of fluorophore to antibodies can be raised by 100–1,000 times, thus achieving detection sensitivities of several orders of magnitude [28].

Another class of fluorescent labels is the lanthanide complexes. Whereas most fluoroscent compounds have decay times of 5–100 nanoseconds, lanthanide fluorophores have much longer decay times. They are not as bright as many other labels, but they have very large Stoke's shifts. Their long lifetimes permit sensitive assays to be developed based on time-resolved methods. Europium complexes excite in near UV and emit at 600–700 nm. Lifetimes are in the 500–2,000 microsecond range. Terbium complexes also fall into this category.

Ruthenium complexes excite in the blue range and emit in the red range (600–700 nm). Lifetimes are typically hundreds of nanoseconds. Proteins can be conjugated covalently to these commercially available metal chelates.

5.6 Quantum Dots

An emerging class of fluorescent labels for lateral flow immunoassays is quantum dots [29, 30]. Quantum dots are nanocrystals or semiconductor nanocrystals that are composed of CdSe, CdS, ZnSe, InP, or InAs, or a layer of AnS or CdS on a CdSe core. When a semiconductor absorbs a photon having energy greater than its bandgap, an electron is promoted from the valence band into the conduction band, leaving behind a positively charged hole. The electron–hole pair is called an exciton. Excitons are like artificial atoms having radii of 1–10 nm, depending on the properties of the semiconductor. As the size of the semiconductor crystal becomes similar to the size of the exciton, strong quantum confinement modifies the exciton properties. With decreasing crystal size, the exciton behaves more like a particle-in-a box. Its energy levels are determined largely by the size of the particle (box) instead of the properties of the bulk semiconductor. Semiconductor nanocrystals that exhibit strong quantum confinement in all three dimensions are called quantum dots. Recombination of the electron and the hole produces light. The wavelength of light is largely determined by the size of the quantum dot. Quantum dots of extremely uniform size can be made and they have a narrow emission bandwidth in the range of 10–50 nm.

Some of the desirable characteristics of quantum dots include:

i) Highly fluorescent with a quantum efficiency on the order of 0.5.
ii) Resistant to photobleaching.
iii) Can be produced in colloidal suspensions with a narrow band emission spectrum.
iv) During fabrication, the diameter of the quantum dots can be selected to achieve emission fluorescence in a variety of colors. This can also be achieved by changing the composition of the nanocrystal (e.g., CdS emits blue light, and InP emits red light) [31–33].
v) Easily amenable to multiplexing both spatially and spectrally.
vi) The luminescent lifetimes of quantum dots tend to be in the range of 30–100 nanoseconds. This is much longer than the background fluorescence and Raman scattering of most sample matrices. One can therefore use time-gated detection to selectively reduce or remove background fluorescence.
vii) Quantum dots have broad absorption bands that extend well into the ultraviolet region. Their emission wavelength is essentially independent of the excitation wavelength, so quantum dots having narrow bandwidth

emission at wavelengths throughout the visible spectrum can be excited by a single excitation wavelength, wavelength segment of spectrum, or source.

viii) Quantum dots are commercially available (e.g., Quantum Dot Corporation, Hayward, CA).

The use of spectral filtering at the detection end adds an additional dimension to spatially multiplexed assays using fluorescent labels, including Quantum Dots. For example, when we use different labels with different excitation and emission spectral to label each antibody to detect different analytes along different capture zones on a membrane [34, 35], without the benefit of spectral filtering, at each capture zone, the noise would consist of the total from each antibody's non-specific binding: the larger the number of analytes and hence antibodies, the bigger the noise. If we apply spectral filtering, the noise comes only from the non-specific binding of the one antibody with the label that we are viewing, at any one capture zone. Therefore, the signal-to-noise ratio can be significantly improved by spectral filtering techniques, allowing for a larger number of analytes to be effectively assayed.

With all these advantages, we can expect to see more lateral flow immunoassays using Quantum dots.

5.7 Upconverting Phosphors

Phosphorescence is the luminescence produced by certain substances after absorbing radiant energy or other types of energy. Phosphorescence is distinguished from fluorescence in that the light emission continues even after the radiation causing it has ceased. The decay time of emission of phosphorescence light is longer (10^{-4} to 10^{2} s) than the decay time of fluorescence emission. Decay times are expressed in a time range of several orders of magnitude and vary with the molecule and its solution environment. Phosphorescence also shows a larger shift in emitted light wavelength than does fluorescence.

The use of upconverting phosphors (UP) was first reported in cell biology by Zijlmans et al. [36] and in immunoassays by Niedbala et al. [37]. The label is a nanoparticle of 200–400 nm in diameter. It consists of a crystalline lanthanide oxysulfide. Antibodies are covalently conjugated. UP absorbs two or more photons of infrared light (usually at 980 nm) and emits light at a shorter wavelength, resulting in what is called an anti-Stoke's shift. This upconverting phosphorescence is not influenced by reaction conditions including temperature or buffer and, since there is no upconverted signal from biological components in the sample, the background signal is low. Multiplexing is possible because different types of particle produce different wavelengths of phosphorescence (e.g., yttrium/erbium oxysulfides are green at 550 nm and yttrium/thulium oxysulfide particles are blue at 475 nm). A commercial product for drug–of-abuse detection had been available [38, 39].

5.8 Bioluminescent Markers

A promising bioluminescent label is the recombinant aequorin available from Sealite Sciences (Bogart, GA). This protein was originally extracted from the jellyfish Aequorea. Upon activation by divalent calcium ions, aequorin emits light as a result of an intermolecular reaction in which coelenterazine (an imidazopyrazine compound bound noncovalently to the protein) is oxidized to coelenteramide, thereby yielding a blue light having a maximum wavelength of about 470 nm and carbon dioxide. The wavelength of light emitted by aequorin is generally in the range of 440–475 nm. The excited state of coelenteramide bound to the protein is the emitter in the reaction. The amount of light generated is proportional to the amount of photoprotein activated [40]. Coelenteramide analogs are commercially available (e.g., Molecular Biosciences Inc. in Boulder, CO).

Recombinant aequorin can be conjugated to antibodies by the introduction of sulfhydryl groups to aequorin, using reagents such as iminothiolane, and then reacts with SMCC-modified antibodies. Iminothiolane (Traut's reagent) is selected because it was found not to interfere with the photolytic activity of aequorin [41, 42].

A lateral flow assay using aequorin was described by Liotta et al. from Immunomatrix, Inc. (Washington, DC) [43], in which aequorin is conjugated to an analyte specific antibody and dried onto a conjugate zone. When sample liquid containing the analyte passes through the conjugate zone, it reconstitutes and reacts with the aequorin-conjugate. The sample and the conjugate then pass through a capture zone on the membrane immobilized with the analyte. Any "free" aequorin-conjugate that has not had its binding sites occupied by the sample analyte is captured by the immobilized analyte. The aequorin-conjugate that has reacted with the sample analyte will pass unhindered through the capture zone and proceeds to a third zone which contains immobilized/caged calcium ions. At this third zone, calcium activates the aequorin and emits light, which is detected by a photodetector located adjacent to this zone.

5.8.1 Luminescent Oxygen Channeling Immunoassay (LOCI)

Homogeneous sandwich immunoassay in which an antigen links an antibody-coated sensitizer dye-loaded particle (250-nm diameter) and an antibody-coated particle (250 nm diameter) loaded with a mixture of a precursor of a chemiluminescent compound and a fluorophore. Irradiation produces singlet oxygen at the surface of the sensitizer dye-loaded particle. This diffuses ("channels") to the other particle held in close proximity by the immunochemical reaction between the antigen and the antibodies on the particles. The singlet oxygen reacts with the chemiluninescent compound precursor in the particle to form a chemiluninescent dioxane, which then decomposes to emit light via a fluorophore-sensitized mechanism. No signal is obtained from precursor fluorophore-loaded particles that are not linked via immunological reaction with an antigen.

This method is similar to the enzyme channeling method described in the section on enzyme labels, but results in a luminescent signal that can be determined by a photosensor and is amenable to use for lateral flow immunoassays [44].

5.8.2 *Advantages and Limitations of Luminescent Measurements*

The main advantages of luminescent (light emitting) labels over fluorescent labels is that fluorescent labels need a light source in addition to a photodetector for its measurement, whereas luminescent labels need only a photodetector. This simplifies the design of the instrument and makes it less expensive.

Light leakage, light piping, and high background luminescence from assay reagents and reaction vessels (e.g., exposing plastic tube to light) are the common factors that degrade analytical performance. The extreme sensitivity of luminescent assays requires stringent controls on the purity of reagents and the solvents (e.g., water) used to prepare reagent solutions. Efficient capture of the light emission from reactions that produce a flash of light (in chemiluminescent assays) requires an efficient injector which provides adequate mixing when the triggering reagent is added to the reaction vessel. Luminescent assays have a wide linear range, but very high-intensity light emission can lead to pulse pile-up in photomultiplier tubes, leading to a serious underestimation of the true light emission intensity.

5.9 Enzyme Labels

Enzyme labels have been used extensively in immunoassays such as microtiter ELISA, chemiluminescent assays, and flow-through (or vertical flow) membrane immunoassays. A primary advantage for using enzyme labels is the potential amplification that can be achieved. However, they are less popular in lateral flow immunoassays (especially self-contained assays) for several reasons:

i) Room temperature storage for extended periods are difficult with enzyme conjugates.
ii) There is a scarcity of stable substrates for commonly used enzyme labels.
iii) Enzyme immunoassays generally require washing steps to remove excess conjugates and development procedures to visualize.
iv) Products of the enzymatic reaction diffuse rapidly away from the label, making localization difficult.

A classic example of a lateral flow immunoassay using enzymes is the enzyme channeling method described by Litman et al. [45, 46], in which a binding event brings two enzymes together at a surface. One enzyme (glucose oxidase) produces a product (hydrogen peroxide) that is a substrate for another enzyme (horseradish peroxidase). Because of the slow diffusion at the surface, the hydrogen peroxide produced has a higher local concentration than in the bulk solution and selectively reacts with the proximal second enzyme. Any of the hydrogen peroxide generated by the first enzyme in the bulk solution is scavenged by the catalase present in the reaction mix.

5.10 Paramagnetic Particles

Paramagnetic particles are colloidal particles of iron oxide, which are only magnetic when placed in a magnetic field so they do not behave as individual dipole magnets that would lead to aggregation and destabilization of the colloidal structure. These colloids can be polymer-coated, enabling the adsorption or covalent linking of antibodies and antigens. The conjugated particles can be used in lateral flow immunoassays in the same way as latex or colloidal gold conjugates. Their detection and quantitation are done by measuring the magnetic flux generated when placed inside a magnetic field.

Quantum Design Corporation (San Diego, CA) has designed an instrument that is able to quantify the amount of paramagnetic particles at different zones on a membrane [47]. This is done by exposing a membrane to a group of small rectangular coils residing between two ferrite E-shaped core magnets. The outer legs of the magnets are longer than the central leg, creating a gap when placed face to face. When the coils are energized, a strong homogeneous magnetic field in the center portion results. A motor is used to align and position the test strip in the center portion of the magnet to completely magnetize the particles for detection by the magnetic field sensors or coils. The sensors are arranged to function as a gradiometer to determine both the position and the amplitude of the magnetic signals. The signals are then amplified and processed to give a numeric value that indicates the quantity of magnetic particles in the analytical region of the membrane.

In a conventional lateral flow immunoassay, the control and test lines are detected visually or by an optical instrument that measures reflectance, contrast, color change, or fluorescence. Such observations of the accumulation of particles at the test line measure at most the top 10 μm of the substrate membrane. Typically, the membranes that are used are a few hundred microns thick. Since the particle-labeled analyte travels not only on the surface but also within the membrane, up to 90% of the analyte is not detected by a measuring device, the eye, or an optical spectrometer. Additionally, optical measurement of analytes in biological fluids can also be difficult due to interfering substances, specular reflection, scattering, self-absorption, and quenching of signal (in fluorescence-based tests). The magnetic reader determines the entire amount of paramagnetic particles in the analytical zone of the membrane, including those present beneath the surface and which cannot be seen optically. The strength of the magnetic signal increases directly with the mass of the magnetic label, and this inherent linearity contributes to its sensitivity, accuracy, and wide dynamic range.

The particle label used in this application must have a number of clearly defined characteristics. It must have negligible magnetic memory to avoid destabilization of the conjugate. The particle size distribution must be small, and the Fe_3O_4 content must be consistent and quantified to ensure signal reproducibility, and the ratio of Fe_3O_4 to matrix material influences both sensitivity of the assay and the separation technique used during conjugation.

5.11 Latex Particles

Latex particles, the first labels used for lateral flow immunoassay, remain as one of the most versatile labels. They are available from many commercial sources at a relatively low cost. Well-documented chemistries are available for conjugation, including simple adsorption and covalent coupling through the amino, carboxyl, and thiol groups. They can be made to incorporate color dyes including fluorescent dyes, and paramagnetic media. Essentially, they can be substituted for any of the labels discussed previously. Fluorescent microspheres, for example, are used more often today than direct labeling of antibodies for membrane-based assays, since they provide the option of non-covalent adsorption and cause less physical modification of the biological. In addition, some fluorescent dyes, which are difficult to conjugate because they do not have reactive chemical groups and those that would lose a significant amount of fluorescence on chemical modification, can simply be incorporated or encapsulated into latex spheres. In general, it is also easier to process and control the F:P ratio with latex beads.

Another significant advantage with latex is that they can be manufactured to incorporate several characteristics together, making them amenable to detection by multiple methods (e.g., paramagnetic particles that are fluorescent and also colored). Another advantage is that latex beads can be made with multiple surface chemistries with controlled molar equivalencies. This allows for different multiple ligands to be specifically and quantitatively conjugated covalently using selective heterobinfunctional linkers. Together, these features render the latex particles to more easily accommodate coupled assays requiring multiple components, such as luminescent-fluorescent energy transfer systems, channeling methods, and multi-point binding assays.

References

1. Edwards, K.A. and Baeumner, A.J. (2006) Analysis of liposomes. J. Talanta. 68(5):1432–1441.
2. Geck, P. (1971) India-Ink immuno-reaction for the rapid detection of pnteric Pathogens. Acta Microbiol. Acad. Sci. Hung. 18:191–196.

3. Waller, T. (1977) The India-ink immunoreaction: a method for the rapid diagnosis of ncehalitozoonosis. Lab. Anim. 11:93–97.
4. Berguist, N.R. and Waller, T. (1983) tA novel simple immunoassay for rapid detection of human IgG antibodies to Toxoplasma gondii. J. Immunol. Meth. 61:339–344.
5. Van Amerongen, A., Wichers, J.H., Berendsen, L.B., Timmermans, A.J., Keizer, G.D., van Doorn, A.W., Bantjes, A. and van Gelder, W.M. (1993) Colloidal carbon particles of a new label for rapid immunochemical test methods: quantitative computer image analyses of results. J. Biotechnol. 30(2):185–195.
6. Bowers, J.S. and Bernards, R.F. (2003) Method to improve the stability of dispersions of carbon. U.S. patent 5641689.
7. Chandler, J., Gurmin, T. and Robinson, N. (2000) The place of Gold in Rapid Tests. IVD Technology 6:37–49.
8. Faulk, W.P. and Taylor, G.M. (1971) An immunocolloid method for the electron microscope. Immunochemistry 8:1081.
9. Frens, G. (1973) Controlled nucleation for the regulation of the particle size in monodisperse gold solutions. Nature (London) Phy. Sci. 241:20.
10. Leuvering, J.H.W., Thal, P.J.H.M., Van der Waart, M. and Shuurs, A.H.W.M. (1981) A sol particle agglutination assay for human chorionic gonadotrophin. J. Immunol. Meth. 45:183–194.
11. Brada, D. and Roth, J. (1984) "Golden Blot" – detection of polyclonal and monoclonal antibodies bound to antigens on nitrocellulose by protein A-gold complexes. Anal. Biochem. 142:79–83.
12. Moeremans, M., Daneels, G., Van Dijck, A., Langanger, G. and De Mey, J. (1984) Sensitive visualization of antigen-antibody reactions in dot and blot immune overlay assays with immunogold and immunogold/silver staining. J. Immunol. Meth. 74:353–360.
13. Handley, D.A. (1989) Methods for Synthesis of Colloidal Gold. In "Colloidal Gold: Principles, Methods, and Applications", Hyat, M.A., Ed. Academic Press, Inc. San Diego, Vol. 1.
14. Goodman, S.L., Hodges, G.M., Tredjdosiewicz, L.K. and Livingston, D.C. (1981) Colloidal gold markers and probes for routine application in microscopy. J. Microsc. (Oxford) 123:201.
15. Geoghegan, W.D. and Ackerman, G.A. (1977) Adsorption of horseradish peroxidase, ovomucoid and anti-immunoglobulin to colloidal gold for the indirect detection of concanavalin A, wheat germ agglutinin and goat anti-human immunoglobulin G on cell surfaces at the electron microscopic level: a new method, theory and application. J. Histochem. Cytochem. 25:1187.
16. Geoghegan, W.D. (1988) The effect of three variables on adsorption of rabbit IgG to colloidal gold. J. Hisochem. Cytochem. 36:401.
17. Horrisberger, M., Rosset, J. and Bauer, H. (1975) Colloidal gold granules a s markers for cell surface receptors in scanning electron microscopy. Experimentia 31:1147.
18. Jones, K. (1999) Investigation of use of conjugate release material for lateral flow immunoassay. Poster Presentation AACC Annual Meeting, New Orleans.
19. Millipore Corp. Bedford MA. (2002) Literature No. TB500EN00. Rapid lateral flow test Streps: Considerations for product development.
20. O'Farrell, B. and Bauer, J. (2006) Developing highly sensitive, more reproducible lateral-flow assays, Part 1: New approaches to old problems. IVD Technology, June issue, p. 41.
21. Danscher, G. (1984) Autometallography: a new technique for light and electron microscopical visualization of metals in biological tissue (gold, silver, metal sulphides and metal selenides). Histochemistry 81:331.
22. Danscher G. (1981) Localization of gold in biological tissue: a photochemical method for light and electron microscopy. Histochemistry 71:81.

23. Holgate, C.S., Jackson, P., Cowen, P.N. and Bird, C.C. (1983) Immunogold-silver staining: new method of immunostaining with enhanced sensitivity. J. Histochem. Cytochem. 31:938.
24. Hacker, G.W., Polak, J.M., Springall, D.R., Tang, S.K., Van Noorden, S., Lackie, P., Grimelius, L. and Adam, H. (1985) Immunogold-silver staining (IGSS) – a review. Mikroskopie (Vienna) 42:318.
25. Horton, J.K., Swinburne, S. and O'Sullivan, M.J. (1991) A novel, rapid single-step immunochromatographic procedure for the detection of mouse immunoglobulin. J. Immunol. Meth. 140:131–134.
26. Haugland, R.P. (ed.) (2002) Handbook of Fluorescent Probes and Research Products, 9th edition. Molecular Probes, Inc. Eugene Oregon.
27. Eberhard, K., Gruler, R., Eberhard, M. and Haberstroh, K. (2007) Developing rapid POC Systems, Part 1: Devices and applications for lateral flow immunodiagnostics. IVD Technology, July/August issue, p. 47.
28. Bonenberger, J. and Doumanas, M. (2006) Overcoming sensitivity limitations of lateral-flow immunoassays with a novel labeling technique. IVD Technology, May issue, pp. 41–46.
29. Chan, W.C.W. and Nie, S. (1998) Quantum dot bioconjugates for ultrasensitive non-isotopic detection. Science 281(5385):2016–2018.
30. Murphy, C.J. and Coffer, J.L. (2002) Quantum dots: a primer. Appl. Spectrosc. 56:16A.
31. Rosenthal, S.J. (2001) Bar coding biomolecules with fluorescent nanocrystals. Nat. Biotechnol. 19:621–622.
32. Han, M., Gao, X., Su, J.Z. and Nie, S. (2002) Quantum-dot-tagged microbeads for multiplexed optical coding of biomolecules. Nat. Biotechnol. 19:631–635.
33. Han, M., Gao, X., Su, J.Z. and Nie, S. (2001) Quantum-dot-tagged microbeads for multiplexed optical coding biomolecules. Nat. Biotechnol. 19:631–635.
34. Chan, W.C., Maxwell, D.J., Gao, X., Bailey, R.E., Han, M. and Nie, S. (2002) Luminescent quantum dots for multiplexed biological detection and imaging. Curr. Opin. Biotechnol. 13:40–46.
35. Lambert, J.L. and Fisher, A.M. (2006) Diagnostic assays including multiplexed lateral flow immunoassays with quantum dots. Patent WO/2006/071247.
36. Zijlmans, H.J.M.A.A., Bonnet, J., Burton, J., Kardos, K., Vail, T., Niedbala, R.S. and Tanke, H.J. (1999) Detection of cell and tissue surface antigens using up-converting phosphors: a new reporter technology. Anal. Biochem. 267:30–36.
37. Niedbala, R., Feindt, H., Kardos, K., Vail, T., Burton, J., Bielska, B., Li, S., Milunic, D., Bourdelle, P. and Vallejo, R. (2001) Detection of analytes by immunoassay using up-converting phosphor technology. Anal. Biochem. 293:22–30.
38. Hampl, J., Hall, M., Mufti. N., Yao, Y.-M., MacQueen, D., Wright, W. and Cooper, D. (2001) Upconverting phosphor reporters in immunochromatographic assays. Anal. Biochem. 288(2):176–187.
39. Steinmeyer, S., Polzius, R. and Manns, A. (2005) Drager Drug Test – Test for illegal drugs in oral fluid samples. In "Forensic Science and Medicine: Drugs of Abuse: Body Fluid Testing", Wong, R.C. and Tse, H.Y., Eds. Humana Press Inc., Publisher, Totowa, NJ.
40. Shimomura, O. and Johnson, F.H. (1978) Peroxidized coelenterazine, the active group in the photoprotein aequorin. Proc. Natl. Acad. Sci. USA 75(6):2611–2615.
41. Stults, N.L., Stocks, N.F., Rivera, H., Gray, J., McCann, R.O., O'Kane, D., Cummings, R.D., Cormier, M.J. and Smith, D.F. (1992) Use of recombinant biotinylated aequorin in microtiter and membrane-based assays: purification of cecombinant apoaequorin from E.coli. Biochemistry 31:1433–1442.
42. Rigl, C.T., Atel, M.T., Rivera, H.N., Stults, N.L. and Smith D.F. (1994) A bioluminescent immunoassay based on recombinant photoprotein. In "Bioluminescence and Chemiluminescence: Fundamentals and Applied Aspects", Campbell, A.K., Kricka, L.J. and Stanley, P.E., Eds. Wiley, Chichester, pp. 345–348.

43. Liotta, L.A., Christiansen, B.C., Day, A.R., Harlacher, T. and Paweletz, K. (1999) Light emitting immunoassay. US patent 5942407.
44. Ullman, E.F., Kirakossian, H., Sing, S., Irvin, B.R., Irvine, J.D. and Wagner, D.B. (1994) Luminescent oxygen channeling immunoassay (LOCI) for human thyroid stimulating hormone. In "Bioluminecence and Chemiluminescence: Fundamentals and Applied Aspects". Campbell, A.K., Kricka, L.J. and Stanley, P.E., Eds. Wiley, Chichester, pp. 6–19.
45. Litman, D.J., Hanlon, T.M. and Ullman, E.F. (1980) Enzyme channeling immunoassay: a new homogeneous enzyme immunoassay technique. Anal. Biochem. 106:223–229.
46. Zuk, F.F., Ginsberg, V.K., Houts, T., Rabbie, J., Merrick, H., Ullman, E., Fischer, M.M., Sizto, C.C., Stixo, S.N. and Litman, D.J. (1985) Enzyme immunochromatography – a quantitative immunoassay requiring no instrumentation. Clin. Chem. 31(7):1144–1150.
47. LaBorde, R.T. and O'Farrell, B. (2002) Paramagnetic-particle detection in lateral-flow assays. IVD Technology, April issue, p. 36.

Chapter 6
Nitrocellulose Membranes for Lateral Flow Immunoassays: A Technical Treatise

Michael A. Mansfield

6.1 Introduction

Lateral flow tests on the market today come in a variety of configurations. In simplest form, the test strip comprises several porous materials mounted on an adhesive backing and covered with an adhesive tape. In more complex designs, the strips are placed into plastic housings. The housing is used to expose the sample pad, maintain proper alignment of the materials, and indicate positions of the test and control lines. While an adhesive backing is normally used for mounting of the porous materials during the manufacturing process, the tape used to cover the materials may be reduced or eliminated depending on the internal design of the housing. In the most complex designs, the housing is designed to hold the materials in the desired alignment. The housing, the materials, and the manufacturing process are integrated so that alignment does not require adhesive materials. Also, a desiccant tablet is often placed inside the housing. A recently introduced pregnancy test even includes on-board electronics that produce the word "pregnant" on a visual display instead of requiring the user to look for a positive test line.

Regardless of the complexity of the test strip, nitrocellulose membranes are common to all lateral flow immunoassay tests. For several reasons, the general perception is that the nitrocellulose membrane is the most critical part of a lateral flow test [1–3]. First, it is the surface upon which the critical immunocomplexes form. Second, it is the surface from which the signal is detected, either visually or electronically. Third, historically, it has been the most difficult material to manufacture consistently. While it is true that the membrane is critical for the formation of the immunocomplexes, overall functionality of the test strip depends on all of the materials, chemistries, design elements, and manufacturing processes.

M.A. Mansfield (✉)
Diagnostic Business Unit, Millipore Corporation, Bedford, MA
e-mail: Michael_mansfield@millipore.com

R.C. Wong, H.Y. Tse (eds.), *Lateral Flow Immunoassay*,
DOI 10.1007/978-1-59745-240-3_6, © Humana Press, New York, NY 2009

This chapter discusses the utilization of nitrocellulose membranes in lateral flow immunoassay test strips. Consideration is given to the manufacturing and testing of the membrane, as well as its physical and chemical properties. Integration of the membrane into finished test strips is discussed relative to the optimization of liquid flow through all of the porous materials to ensure uniform signal development at the test and control lines.

6.2 Historical Perspective

Nitrocellulose membranes have been manufactured for filtration purposes for many decades [4]. In the 1970s, molecular biologists discovered the utility of nitrocellulose membranes as a substrate for molecular detection [5–7]. Building on this experience, lateral flow tests were patented and introduced in the 1980s [8–10]. The primary change with the lateral flow format was the mode of exposure of the probe molecules to the bound biomolecules. Rather than exposing the surface of the membrane to the probe molecule in a comparatively large volume of buffer [5–7], the probe molecule was carried through the pores parallel to the plane of the membrane as liquid moved from one end of the membrane to the other [8–10]. To facilitate liquid flow, membranes had to have a nominal pore rating of >3 μm. Membranes with pores size ≤ 0.45 μm, which were used in molecular biology applications, did not have lateral flow rates sufficiently fast to be of practical use. Fortunately, membrane manufacturers were already producing membranes with pore ratings up to 8 μm. Problems arose with the consistency of membrane performance, however, because these membranes were qualified for normal flow filtration and not for lateral flow tests. Consequently, manufacturers had to develop release criteria related to such use. Requirements for faster-flowing membranes also caused manufacturers to develop membranes with pore sizes estimated to be 15–20 μm. Because the nominal pore size parallel to the plane of the membrane cannot be measured, lateral flow membranes are classified on the basis of the lateral flow times (Table 6.1; also see Section 6.4.2 and Ref. 2).

Table 6.1 Lateral flow membranes manufactured by Millipore Corporation

Designation	Flow time (s/4 cm)	Relative flow rate	Relative sensitivity
Hi-Flow Plus 240	240	Slow	High
Hi-Flow Plus 180	180	↓	↓
Hi-Flow Plus 135	135		
Hi-Flow Plus 120	120		
Hi-Flow Plus 90	90		
Hi-Flow Plus 75	75	Fast	Low

6.3 Membrane Manufacture

6.3.1 Raw Materials

Nitrocellulose membranes are produced by slow and controlled precipitation of polymer from a solvent system [3, 11]. The manufacturing process begins by preparing a lacquer, which is a proprietary combination of nitrocellulose polymers and a defined solvent system. While the components of the solvent system are easily controlled for consistency, the polymer represents a significant challenge. Nitrocellulose is manufactured as an industrial commodity, with only a small amount of total production being converted to membranes. The polymer is typically characterized on the basis of solution viscosity at defined concentrations and related back to mean molecular weight. While this measurement indicates the bulk properties of a given batch of nitrocellulose, blending different grades of polymers to produce a desired viscosity leads to variation in the molecular weight distribution between lots. This variation, affecting both the dissolution properties when the lacquer is prepared and the precipitation characteristics when the membrane is cast, has to be manageable within the process controls of the casting equipment.

Surfactant is another key raw material [1–3]. Nitrocellulose is naturally hydrophobic, and membranes made from nitrocellulose do not wet out in water. To produce membranes that are wettable, manufacturers add a surfactant as a component of the lacquer or apply it to the membrane at the end of the casting process. Manufacturers use different surfactants, and the final concentrations in the membranes vary. The surfactants used are proprietary, but they have been screened for general compatibility with the antibody systems used in lateral flow tests. Still, experience has demonstrated that membranes with similar flow times but from different manufacturers need to be tested empirically for compatibility with specific test chemistries.

6.3.2 Membrane Casting

The primary goal of the casting process is to produce a membrane that is consistent for structure, thickness, and capillary flow time [2]. This is particularly challenging considering that the lacquer is applied to a moving belt in a layer that is a fraction of a millimeter thick. As the belt carries the lacquer through the casting machine, the solvents are evaporated from the membrane under moving air at controlled flow rate, temperature, and humidity. To achieve the large pore sizes typical of lateral flow membranes, evaporation has to be very slow to allow the polymer strands to associate with each other prior to precipitation. Until enough solvent has evaporated to precipitate the polymer, any disturbance to the lacquer can lead to unacceptable thickness variations and structural artifacts. Significant improvements in casting technology specific for these membranes have been made through the years.

Early generations of lateral flow membranes were cast directly onto stainless steel belts and collected at the end of the machine as unbacked membranes. Membranes without a backing are difficult to handle because of their inherent brittleness, and significant yield losses are encountered each time a roll of membrane is unwound and rewound. As membrane manufacturers gained experience in the casting of these membranes, it was discovered that they could be cast directly onto polyester film with sufficient adherence that the nitrocellulose did not easily delaminate [2, 3, 11]. This improved the handling characteristics of the membrane. Fortuitously, the films did not interfere with the performance of the membrane in the finished test strip. Unbacked membranes continue to be produced for test strip by manufacturers whose equipment is optimized for unbacked membranes, but backed membranes are preferred for new tests because of the handling advantages. The thickness of the backing ranges from 2 to 7 mil (50 to 225 μm), depending on the membrane manufacturer.

6.4 Membrane Testing

6.4.1 Sampling Plan

Lateral flow membranes are cast in a continuous, web-based process and collected as master rolls typically 100 m in length. This length has become the preferred length for the test strip manufacturers using reel-to-reel processing equipment. For quality control purposes, a sample is taken between each pair of rolls and tested. This sampling plan produces data sets for the "head end" and "tail end" of each master roll. Both data sets are considered in the assessment of an individual master roll. At each internal sampling point, the data set serves as the "tail end" set for the first roll and the "head end" set for the roll that follows. An example of this sampling plan, as applied by Millipore Corporation, is shown in Fig. 6.1 for capillary flow time. At each location, six coupons are

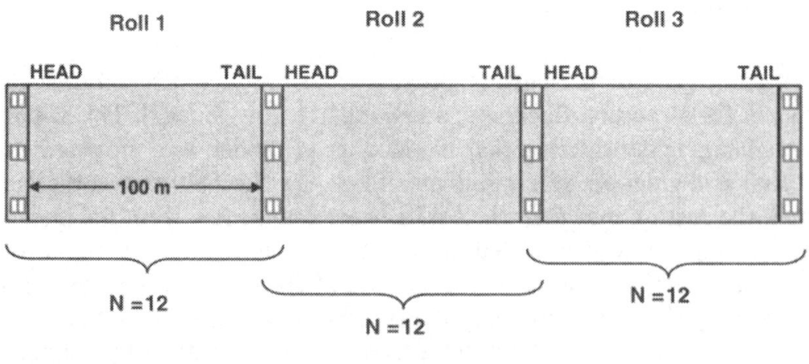

Fig. 6.1 Sampling plan for the evaluation of capillary flow time
The sampling plan used for the evaluation of capillary flow time is presented diagrammatically. Dimensions are not to scale.

tested. The release of the master roll is based on the 12 measurements combined from the head and tail ends. Since one data set applies to two master rolls, a failure within that data set is the cause for rejection of both master rolls. Between sampling points, there is a possibility for membrane to be out of specification. Sampling at 100-m intervals represents a balance between maximizing product yield while minimizing the release of out-of-specification material. Shorter sampling intervals reduce the roll length and increase the cost of testing.

6.4.2 Capillary Flow Time

Lateral flow membranes are evaluated on the basis of capillary flow time (also referred to as wicking time, see Chapter 8), which is the time required for water to travel up and completely fill a 4-cm long strip of membrane [2, 3, 11]. This value is very easy to measure. The selection of 4 cm as the length of the test strip was made for practical reasons. Capillary flow time can be measured on shorter strips. Initially, however, the flow is very rapid. On membranes that flow very fast, it is difficult to call the endpoint of the test consistently, leading to imprecision in the measurement. By extending the length of the strip to 4 cm, the endpoint is more obvious.

Capillary flow time (s/4 cm) is inversely related to capillary flow rate, which is the distance traveled per unit time [2]. Flow rate is difficult to measure reliably, because the flow rate decays exponentially as the water front moves up the membrane. Thus, it is constantly changing. While capillary flow time is measured during membrane testing, it is the flow rate that determines how rapidly a test strip runs and how sensitive the detection system will be. Capillary flow rate is related to the size of the pores parallel to the plane of the membrane. As pore size increases, the flow rate of the membrane increases. For a given membrane type, the membrane manufacturer will specify the capillary flow time. Strip manufacturers should also understand the degree of variability that can be expected within a lot of membrane and the statistical analysis that the membrane manufacturer uses to assess variability. A membrane with highly variable capillary flow time can be very difficult to fabricate into test strips with predictable performance characteristics.

When considering the test strip as a whole, the membrane is the material that normally determines the overall flow rate of the system and the time required to achieve a signal. Development of signals at the test and control lines is a non-equilibrium process, because the analyte and detector particles are being actively carried in the liquid stream and can interact with the capture reagents only for the brief time that they are sufficiently close at the molecular level. As soon as the last detector particles have passed the test line, no further signal development will take place. The effects of membrane flow rate and test line placement on sensitivity have been discussed extensively elsewhere [2, 11].

6.4.3 Membrane Thickness

Membrane thickness is measured using a standard thickness gauge. This parameter is important for several reasons. First, the volume of liquid required to saturate a given area of membrane is determined by the pore volume, which in turn is determined by the thickness of the membrane [2]. For example, when a capture reagent buffer is dispensed onto the membrane, it is likely to spread farther on a thinner membrane. This can affect the width of the capture reagent line, which in turn defines the width of the signal line when the strip is run. Second, dispensing of the capture reagents onto the membrane can be affected by thickness variation [11]. With aerosol application, the cross-sectional area of the stream on the membrane surface can change because the gap between the dispenser tip and the membrane surface varies. With a contact dispenser, variation in the angle between the tip and the membrane surface can affect the consistency of the liquid stream. Third, membrane thickness is important when the strip is placed into a plastic housing [2]. Where the membrane is subject to compression to maintain contact with pad materials, thickness variation can lead to crushing of the membrane and the pads.

It must also be recognized that membrane manufacturers supply membrane within different ranges. Standard nitrocellulose membranes can be as thin as 100 μm and as thick as 150 μm. The range for a specific membrane is determined by the engineering design of the equipment used for its manufacture. When comparing membranes with similar flow times from different manufacturers, consideration must be given to thickness differences for the reasons outlined in the preceding paragraph.

6.4.4 Visual Quality

Visual quality is a subjective assessment of the structural uniformity of the membrane when viewed under various lighting regimes. The membrane should appear uniformly white with no obvious irregularities [2]. Variations in the precipitation of the nitrocellulose are frequently manifested as visual defects in the membrane. When visual defects are extensive, the entire membrane lot may have to be rejected. This typically indicates a fundamental problem with the casting process. If the visual defect is intermittent or infrequent, it may be feasible to cull out the affected area. If an area with a visual defect is encountered during test strip manufacture, it should not be used.

6.5 Membrane Performance

6.5.1 Protein Binding

Protein binding is essential to the function of the membranes in a lateral flow test, but the properties of the nitrocellulose itself are not normally an issue [1–3].

Because nitrocellulose is inherently hydrophobic, it has a high adsorptive capacity for proteins. Lateral flow membranes typically adsorb more than 100 μg of IgG per cm^2. At the concentrations of capture reagents typically applied to the membrane, there is five- to tenfold more binding capacity than necessary. Adsorptive capacity decreases with the molecular weight of the protein. A weak signal is often interpreted as reduced protein binding on the membrane, but this is usually due to solution chemistries that interfere with adsorption to the nitrocellulose or promote desorption when the sample wicks through the test and control lines [11].

To maximize adsorption, antibodies and other proteins should be applied to the membrane in buffers that are preferably free of salt, surfactants, and sugars [2, 11]. The buffer should also be at a low concentration so that crystals dried in the membrane are not of sufficient abundance to occlude the pores. If binding activity of the antibody requires the addition of compounds that might interfere with adsorption, the concentrations used should be no higher than required to maintain antibody functionality.

6.5.2 Blocking

Blocking of the membrane to prevent nonspecific binding of the detector particle and analyte is not absolutely essential to fabrication of a functional lateral flow immunoassay test strip. There are many test strips on the market that do not use a blocking agent; however, blocking agents are required for functionality of some tests because of the nature of the particular sample and antibody system [1, 2, 12]. If a blocking agent is desired, one strategy is to include it as a component of the buffer system dried into the sample pad. The blocking agent dissolves upon addition of the sample and co-migrates with the sample along the strip. From a manufacturing standpoint, this is the simplest approach to adding a blocking agent.

The second option is to apply the blocking agent directly to the membrane by spraying on a defined amount of blocking solution or dipping the membrane into a reservoir of blocking solution. The concentration and type of blocking agent must be determined empirically for compatibility with the sample and antibody system. Care must be taken to avoid excess blocking agent, which can dry down as crystals that occlude the pores. A wash step in buffer alone may be required to remove the excess.

Although not necessarily used for blocking, applying blocking agents can improve the flow characteristics of the membrane. As discussed earlier, surfactants are added to nitrocellulose membranes to make them wettable. While membrane manufacturers test membrane flow with water, test strips are run using a variety of solutions including buffers, urine, saliva, serum, tissue extracts, and environmental samples [13]. These may not have the same flow characteristics as water and often flow at considerably slower rates. Also, application of the capture reagents to the membranes results in zones with a different chemical environment resulting from drying of the buffer salts, capture reagents, and any other additives into the membrane. Application of the capture reagent can also cause redistribution of the surfactant. The sample may not flow through these areas with the same efficiency as through the rest of the membrane. By applying a blocking solution to the membrane, the nitrocellulose becomes uniformly coated with a single chemical species, the blocking agent. This can improve flow consistency through the membrane.

6.5.3 Membrane Handling

From initial manufacturing until completion of test strip fabrication, membranes are processed through multiple pieces of equipment, brought into contact with other materials, and treated with various chemistries. While the variety of processing schemes is too complex to describe comprehensively in this chapter, these general guidelines apply to both manual and automated assembly of test strips.

1. Minimize contact with the surface of the membrane. When a sharp edge comes into contact with the membrane, it produces a dent. Depth depends on the force applied, but in all cases the discontinuity can affect flow consistency and signal uniformity.
2. If a contact dispenser is used to apply capture reagents to the membrane, the tip must not leave a groove in the membrane [11]. Normally, the tip needs to be composed of a flexible plastic. Tip design and material composition need to be matched to the design of the dispensing equipment.
3. Contact with the edge of the membrane roll should be avoided. If the membrane is unbacked, damage at the edge can serve as the starting point for a break across the width of the membrane when tension is applied. If the membrane is backed, nitrocellulose can flake off the edge and contaminate the manufacturing system.
4. Surfaces that come into contact with the membrane, such as rollers, must be kept clean. If a surface is contaminated with debris, an impression of the debris will be made in the membrane when downward force is applied. In reel-to-reel systems, surface defects sometimes appear at a regular interval, corresponding to the diameter of one of the rollers. The defect also appears identical at each occurrence. This indicates that a piece of debris was stuck to the roller when the membrane was processed through the machine. Cleanliness also applies to manufacturing settings where membrane sections are stacked on top of each other.

6.5.4 Membrane Storage

Membrane storage conditions vary depending on the stage of the test strip manufacturing process [1, 2]. Up until the point that reagents are going to be applied, the membrane can be stored under ambient conditions (15–30°C, 20–80% relative humidity). A condensing atmosphere should be avoided as liquid in the pores can cause redistribution of mobile components, such as the surfactant. When the membrane is being prepared for application of the capture reagents, it should be allowed to equilibrate to the humidity of the dispensing room, particularly if the membrane is being brought in from a drier environment. Humidity from the air acts to hydrate the surface of the nitrocellulose and improves the absorption of the capture reagent solutions. Once the capture reagents have been applied, the membrane should be dried completely to produce maximum adsorption of the capture reagents. It should then be stored under desiccation or in a dry room at <15% relative humidity. If the membrane is to be blocked, it will have to be brought back to ambient humidity first, processed through the blocking solution, dried, and returned to dry storage. If possible, assembly of the test strips should take place in a dry room.

6.6 Flow Properties

The flow properties of the membrane related to pore size and surfactant have been discussed previously. There are, however, other aspects of the flow that merit discussion because of their impact on the performance of the test strip and the ability to generate results that are predictable and consistent. This involves not only flow through the membrane but also flow through the other porous materials and the test strip as a whole. Depending on the test design, the ability of the strip to allow particle flow has to be considered (see Section 6.6.2). There are two important aspects to liquid flow. First, liquid flows preferentially along the path of least resistance [11]. Second, it is important to recognize that the membrane cannot compensate for flow problems elsewhere in the test strip [11]. Since the functionality of the test depends directly on liquid flow [1, 2, 11], it is important to understand how to optimize flow within the strip.

6.6.1 Porous Pads

6.6.1.1 Manufacture

Porous materials are used as the sample, conjugate, and absorbent pads [2]. Most commonly, nonwoven materials are used: glass fiber for the conjugate pad and cellulose papers for the sample and absorbent pads. Other materials are sometimes used, including various types of woven fabrics. Porous plastic wicks are an integral part of many urine-based tests, where the user is instructed to place the wick in the urine stream to collect the sample. The porous structures of these materials are quite different from membranes because of differences in the ways they are manufactured. Glass fiber pads and cellulose papers are manufactured by suspending the appropriate fibers in a large quantity of water. The dilute slurry is applied to a rapidly moving, porous screen that permits the removal of the water under vacuum. The fiber mat is processed through various rolling and drying steps to yield the final material. Large master rolls are then processed into narrower widths and shorter lengths as required by test strip manufacturers.

The key attributes of the pad materials are the bed volume and the thickness [2]. The bed volume, defined as the volume of air in the pores per unit surface area (e.g., $\mu L/cm^2$), determines how much volume of sample is required to saturate the structure and what volume of reagents can be dried into the structure during test strip manufacture. The concentration of the sample pad buffer and detector particle solution can be modulated within the constraints of the bed volume specification. The thickness is also important as strips that are placed into plastic housings are under physical constraint (discussed in 6.6.4).

6.6.1.2 Sample Pad

For almost all of the tests currently marketed, the sample enters the test strip through the sample pad. This material is impregnated with buffer salts, surfactants, and other chemical agents that make the sample suitable for interaction with the detection system. If the analyte of interest is present in the sample, it must be capable of binding to the capture reagents on the detector particles and the membrane. The purpose of the sample pad is to modulate any chemical variability in the sample so that the signal produced is proportional to the concentration of analyte.

6.6.1.3 Conjugate Pad

From the sample pad, liquid is transferred onto the conjugate pad. The primary function of the conjugate pad is to hold the detector particles in a dry state so that they are functionally stable until resuspended by the sample [2]. As the liquid moves through the glass fibers, the detector particles need to be released rapidly and quantitatively, as well as consistently between individual test strips. Every effort should be made to ensure that the distribution of detector particles in the conjugate pad is uniform. The consistency of the signals at the test and control lines cannot be any better than the uniformity of distribution on the conjugate pad. Movement of particles in the membrane is dictated almost completely by the direction of liquid flow.

One artifact commonly seen is channeling of the detector particles on the membrane. Although this is often attributed to a problem with the membrane, it is actually a problem with the transfer of liquid from the conjugate pad onto the membrane. If the flow is channeled through different parts of the conjugate pad, either due to non-uniform transfer from the sample pad, highly variable pore diameters in the conjugate pad, or non-uniform transfer to the membrane, the detector particles move into areas where the flow is fastest. This appears as streaks of detector particles on the membrane. In extreme cases it can result in patchy color development at the test and control lines. Darker zones are coincident with the streaks; lighter zones are coincident with areas between the streaks. Avoiding this artifact requires selection of pad materials with minimal variability in fiber density and distribution.

6.6.1.4 Absorbent Pad

The only function of the absorbent pad is to serve as a sink for the liquid that is processed through the strip [2]. The key attribute of the absorbent pad is the bed volume. There has to be enough pore volume in the absorbent pad to accommodate the full volume of sample that needs to be processed. Once the absorbent pad is full, liquid will stop flowing through the strip.

After a test strip has been run and the results noted, it should be discarded. If a visual record of the result needs to be retained, an electronic image should be made. Alternatively, the absorbent pad and the other pad materials can be removed from the strip and the membrane archived directly. Used test strips should not be stored with the pads retained on the test strip. With prolonged standing, liquid evaporates from the exposed surfaces of the porous materials. The absorbent pad then serves as a reservoir, leading to backflow of liquid onto the membrane. This backflow carries excess detector particles back onto the strip, and can lead to nonspecific and misleading signal development at the test line. This is particularly problematic for samples considered to be negative for the target analyte.

6.6.1.5 Pad Configurations

Another aspect of the liquid flow in the pads is the degree of overlap that they have with each other and the membrane. As the sample flows through the pads, it resuspends the dried chemistries and carries them onto the membrane [1–2]. While each pad fills with liquid, this does not necessarily mean that there is actually flow in any specific region. An obvious example of this is the upstream region of the sample pad in a strip placed into a housing that has a discrete well for sample application. This upstream area wets out when the sample is applied, but does not contribute anything to the development of the signal because the primary flow path is in the downstream direction toward the membrane. Where the pads overlap, it is assumed that the liquid flows through the entire volume of the material. If the overlaps are not configured properly, though, dead spaces can occur where there is little or no flow of the sample. The strip design should allow for complete transfer of the detector particles from the pad. If the flow path has been optimized, there should be no residual color in the conjugate pad after the strip has finished running.

The dimensions of the pads vary between different test strips. Sample pads tend to be much longer than conjugate pads so that they can hold enough dry chemistry to be effective for the entire volume of sample processed through the strip. The conjugate pad tends to be smaller because it holds only a defined amount of detector particle. When these pads are integrated with the membrane, they are arranged in defined positions. For consistent test strip performance, the alignments should be maintained under as tight a specification as practical [12]. It should be obvious that variation in the positioning of the materials changes the degree of overlap and, consequently, the flow properties of the finished strip.

6.6.2 Membrane Flow

Fabrication of the membrane into a finished test strip presents the opportunity to introduce multiple artifacts that adversely affect flow. Figure 6.2 shows different types of flow patterns on the membrane. When the top of the flow front trails behind the bottom, it is often accompanied by unusual wetting patterns and, thus, is visually obvious. When the top edge is leading, this is often difficult to detect unless the separation is greater than 1 mm. Variations in the flow rate at the top and bottom of the membrane indicate problems with the structural uniformity of the pores at the microscopic level or problems with the chemical uniformity of the nitrocellulose surface through the depth of the membrane (Fig. 6.2A). While structural problems relate to membrane manufacturing, chemical uniformity can be related to contaminants that have been introduced onto the membrane surface. Variations in flow patterns can also be seen across the

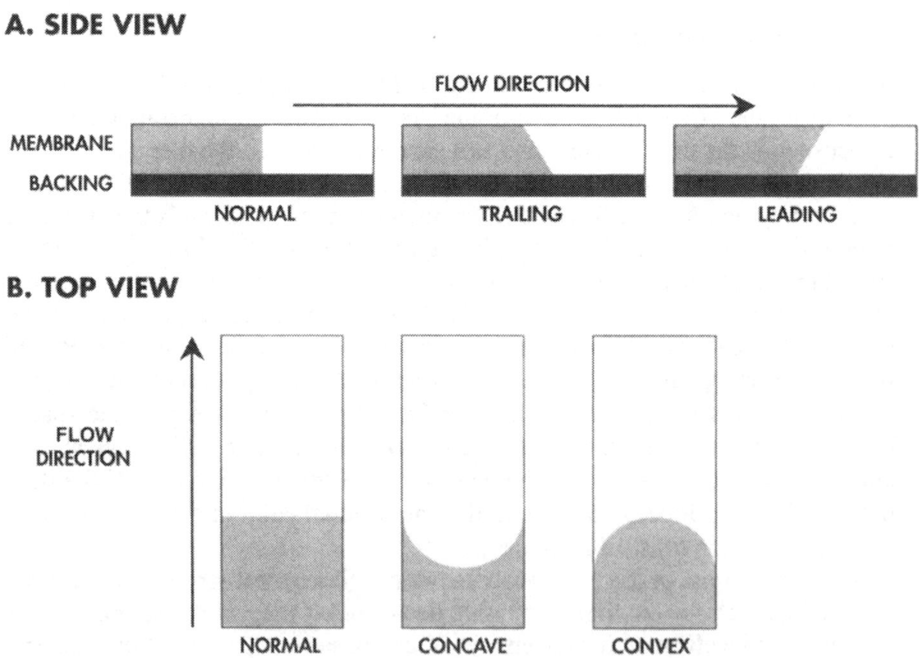

Fig. 6.2 Liquid flow profiles through lateral flow membranes

width of the strip (Fig. 6.2B). If the strip is not placed in a housing, these variations are readily observable. If, however, the strip is inside a housing, they may go unnoticed because the edges of the strip are often obscured by the housing. These types of flow variations are most typically caused by problems with the fabrication process that have resulted in damage to the membrane.

In the finished test strip, every exposed surface of the membrane may have come into contact with another material or been subjected to a mechanical process. Some examples (see Fig. 6.3) are:

a) The leading and trailing edges of the membrane were in contact with metal blades when the membrane roll was originally slit by the manufacturer.
b) The sides of the membrane were in contact with metal blades when the individual strip was cut from the master card.
c) The leading and trailing edges of the membrane are in contact with the conjugate and absorbent pads in the finished strip.
d) Most of the surface of the membrane is in contact with adhesive tape (when included in the strip design).

There are several problems that can arise from this contact. First are problems along the cut edges. The structures of the leading and trailing edges are typically fine, and membrane manufacturers are keenly aware of the need to have cleanly cut edges on their membranes. The most significant problems occur along the sides of the strip. Ideally, the membrane will be in full contact with the adhesive tape on the top and the backing on the bottom, with the nitrocellulose extending fully to the edge (Fig. 6.4). Damage can occur, however, when individual strips are cut because of the range of mechanical properties of the various materials. In some cases, the membrane separates from the adhesive tape. In other cases it separates from the backing. If a channel is opened up (Fig. 6.4), this will provide an unobstructed path for the sample to flow rapidly down the edge, resulting in a concave flow front (Fig. 6.2B). In

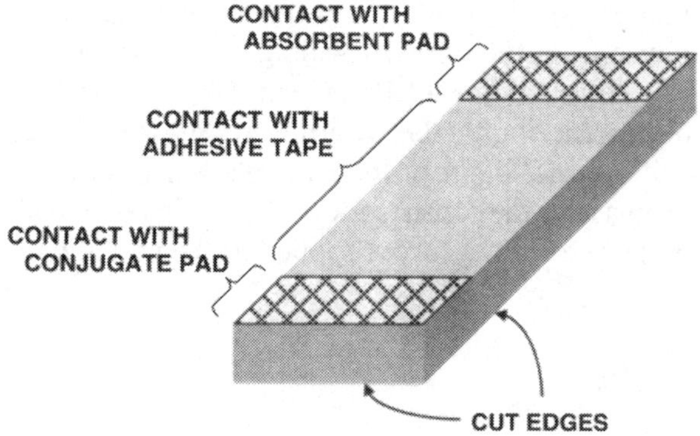

Fig. 6.3 Contact areas on membrane after integration into a finished test strip

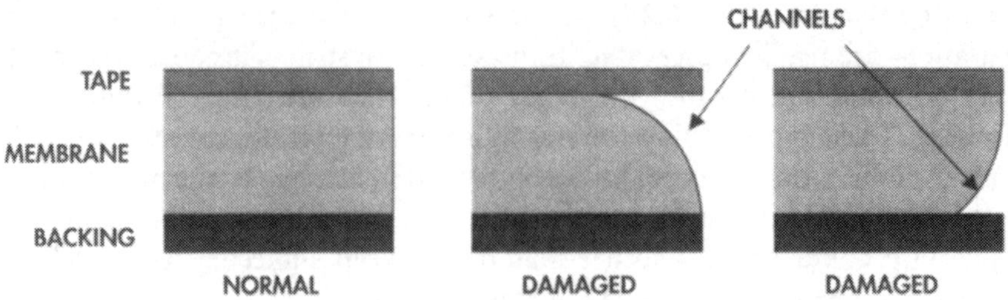

Fig. 6.4 Deformation of membrane edge after cutting
Arrows indicate channels that have opened up as a consequence of the membrane being pulled away from the tape or backing.

other cases, cutting crushes the edges of the strip, reducing the pore size. This causes the sample to flow more slowly along the edges, resulting in a convex flow front (Fig. 6.2B). Even in the absence of adhesive tape, improper cutting can damage the edge of the membrane (Fig. 6.5A).

When the cutting process damages the edges of the membrane, it usually damages the edges of the pad materials, too. Where the conjugate pad and absorbent pad overlap the membrane, their fibrous structure is often pressed down into the surface of the membrane (Fig. 6.5B). This damages the pore structure and can lead to non-uniform transfer of the sample between the pads and the membrane. In extreme cases, the edges of the pads can be so severely crushed that the mechanical force on the pads causes them to bow upward along the central axis and out of contact with the membrane. When this occurs, the primary point of liquid transfer is along the edges. The detector particles take this path, causing the signal lines to be more intense at the edges.

Avoiding these problems requires that the cutting equipment be appropriately designed for the materials involved. Cutting blades should be kept sharp, and the equipment should be kept as clean as possible. The blade(s) also need to be kept free of adhesive. Depending on its flow properties, adhesive can build up on the blades. The adhesive can then stick to the edges of the strip, causing the materials to pull apart as the blade completes the cutting stroke. Excess adhesive can be removed from the blade by wiping the surface with a swab that has been dipped in alcohol. If the blade is removed as part of a maintenance procedure, replacement should be carefully monitored for conformance to mechanical specifications. Slight shifts in the placement of the blade can cause significant problems with the consistency of the cutting due to alterations in the direction and magnitude of the forces applied to the materials.

6.6.3 Strip Width

In the context of the discussion above, it is valid to ask how wide should the strip be. From a cost standpoint, a thinner strip is more economical to

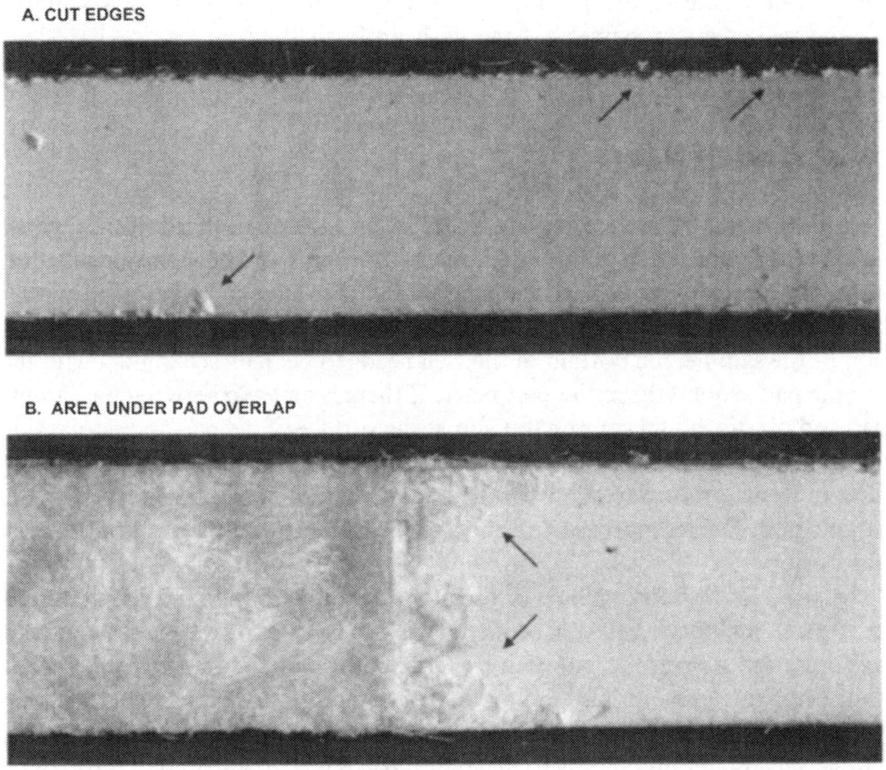

Fig. 6.5 Membrane damage resulting from processing into finished test strips
A. Arrows point to regions with a high degree of damage to the edge resulting from the cutting mechanism.
B. Arrows point to areas of the membrane surface that have been severely damaged by contact with the pad.

produce. However, there are practical limits to how thin a strip should or can be cut. First is the difficulty in reading the signal. Although strong signals are fairly easy to see, the readability of a weak signal depends on the contrast between the signal line and the adjacent membrane area. If there is too little membrane to provide adequate contrast, a weak signal may be missed. Second is a mechanical limitation of the pad materials. Because the pads are typically nonwoven materials, adhesion to the card backing is determined by the small fraction of fibers that are actually in contact with the adhesive. The force of the cutting stroke can easily overwhelm the adhesion, causing the pad material to delaminate and fall apart. Third is the proportion that any edge effects contribute to the overall appearance of the test line. If 0.5 mm on each edge of the strip is subject to aberrant flow, this represents 33% of a strip 3 mm wide. If the strip is 6 mm wide, only 16% of the width is affected. For a quantitative test, it may be desirable to use a wider strip so that the membrane area scanned by the reader encompasses a zone with uniform flow.

6.6.4 Housing Design

Test strip manufacturers often are unskilled in the design of a housing, particularly the features that define and limit the flow path of liquid through the test strip. When a housing is used, the internal features have to be designed so that the flow path is clearly defined. When the housing includes a well for application of the sample, the bottom of the well needs to be in direct contact with the sample pad around the entire perimeter. If there is any gap between the sample pad and plastic, liquid can run out across the surface of the pad and pool inside the housing. The depth of the well and the internal dimensions of the housing have to be determined to allow for full contact without completely crushing the sample pad. Overcompression of the sample pad can limit the rate of sample absorption.

Internal to the housing, bars or pins are often used to hold the pad materials in contact with each other. Consideration needs to be given to the inherent thickness variation of the materials present at the point of compression. If this is done improperly, materials tending to the low end of their thickness specification can lead to a strip that is not thick enough to contact the top of the housing. Conversely, materials tending to the high end can lead to a strip that is too thick and overcompressed when placed in the housing [12].

The shape of the internal features is also important in defining the flow path. When no adhesive tape is present, a compression bar should span the width of the test strip, especially where the conjugate pad overlaps onto the membrane. If the bar is too narrow, liquid can move across the top of the conjugate pad from the sample pad, travel across the top of the conjugate pad around the edges of the bar, and then cascade down the front face of the conjugate pad, where it pools on the surface of the membrane. This opens a flow path for the liquid, which reduces flow through the conjugate pad. If it occurs before all of the conjugate has transferred onto the membrane, sensitivity will be reduced.

The next area of the housing that needs to be considered is the viewing window. The depth of the viewing window needs to be defined so that the bottom edge is not pressed into the membrane. This will alter the flow path through the depth of the membrane. There is no requirement that the bottom of the viewing window be in direct contact with the surface of the test strip. Many test strips enclosed in housings still include an adhesive tape or a piece of clear plastic that serves to protect the membrane from incidental splash when the test strip is run. This also serves to protect the membrane from mechanical damage.

Finally, there is the positioning of the strip in the housing along both dimensions. Many housings include guide bars that determine the longitudinal positioning of the strip so that the test and control lines are properly registered with the viewing window and labels on the exterior of the housing. It is important to recognize that shifts of only a millimeter can have a negative impact on the registration of the pad materials with the compression points within the housing. (The same problem can occur if the pad materials and membrane are not kept in constant alignment when the test strip is assembled.) Lateral positioning of the strip in the housing requires matching the position of the guide bars to the width of the cut strip. This is straightforward, and lateral placement of the strip is not usually a problem. The viewing window is often narrower than the width of the strip so that aberrant flow on the edges is hidden from the user.

The choice of housing needs to be considered early in product development. An off-the-shelf cassette will have predetermined design features that will dictate the dimensions of the materials used in the test strip. In some cases, this constraint may not allow development of an economically viable test strip. Custom-designing a housing allows for greater flexibility in the configuration of the test strip, but there will be additional costs associated with specifying the design, testing prototypes, and committing to manufacture of the final design.

6.6.5 Particle Flow

Beyond the effects of test strip design and membrane processing on liquid flow, consideration needs to be given to the interaction between the membrane and the particles that are present in the flow stream. Particles may be an integral part of the detection system (e.g., colloidal gold, latex beads, magnetic beads) or comprise the analyte of interest (e.g., spores, bacterial cells). The ability of a particle to migrate through the membrane is related to two membrane attributes. First is the pore size. The pores in the membrane have to be sufficiently large to accommodate the particle. From a theoretical standpoint, the slowest flowing membranes with pore sizes estimated at ~3 μm should be able to accommodate most test systems. The second attribute is the flow rate of the membrane. As the diameter of the particle increases, physical resistance to forward movement increases. As the flow rate of the membrane decreases, there is less force from the moving liquid to push the particle forward.

When the pore size and flow rate are considered along with the desire to have tests that finish running in a reasonable amount of time, there are practical limitations on which membranes can be used for different types of particles. For the systems using only colloidal particles, which typically average 40 nm in diameter [13], there are no practical limits to particle flow with the lateral flow membranes on the market today. For larger particles, there is a threshold on flow rate below which particle migration will not occur in a timely manner. With latex and magnetic beads, which can be 0.5 μm or greater in diameter, practical limitations on particle flow restrict their usage to medium- to fast-flowing membranes. Spores and cells have to be evaluated on a case-by-case basis.

Another aspect of particle flow relates to their physical state in solution. Ideally, the particles will be monodisperse regardless of diameter [13]. If the particles associate as dimers, trimers, and larger aggregates, their ability to flow through the membrane will be very limited. For detector particles, aggregation is avoided by applying appropriate chemistries to the test strip. Aggregation of colloidal gold can be detected by a color change from cherry red to purple, blue, or, in extreme cases, clear. For latex and magnetic beads, there is no distinct color change that takes place with aggregation. One place to look for aggregates, however, is at the interface between the conjugate pad and the membrane. When the conjugate pad is peeled away after the test is finished, residual color is sometimes seen on the membrane surface. For biological samples, monodispersion can be more difficult to achieve because of the chemical nature of the particles and the type of sample being processed. Large aggregates are likely to be filtered out by the sample pad and may never reach the membrane to be detected.

6.7 Final Comments

Nitrocellulose membranes are a critical material in lateral flow assays. Membrane performance in a finished test strip depends on the consistency of the membrane from both a structural and chemical standpoint. Optimal performance of the membrane, as determined by the sensitivity and specificity of the test, depends on the efficiency and consistency of immunocomplex formation at the test and control lines. The other materials and chemistries are also critical to membrane performance. In combination with the appropriate manufacturing processes, they need to be optimized to allow for the highest degree of consistency.

6.8 Summary

Nitrocellulose membranes are considered to be the most critical material in lateral flow test strips. The membranes supplied today have standard filtration membranes as progenitors and are challenging to manufacture. Lateral flow membranes across a range of flow speeds and relevant quality control tests have been developed to accommodate the requirements of test developers. Proper utilization of nitrocellulose membranes requires understanding the key physical and chemical properties that relate to test strip functionality. Other important aspects include handling the membrane properly to minimize physical damage and integrating the membrane into the test strip to optimize flow consistency.

References

1. Jones, K. D. (1999) Troubleshooting protein binding in nitrocellulose membranes. IVD Technology 5(2):32–41.
2. Millipore Corporation. (2002) Rapid Lateral Flow Test Strips: Considerations for Product Development. Lit. No. TB500EN00. Bedford, MA.
3. Beer, H. H., Jallerat, E., Pflanz, K., and Klewitz, T. M. (2002) Qualification of cellulose nitrate membranes for lateral-flow assays. IVD Technology 8(1):35–42.
4. Miles, F. D. (1955) Cellulose Nitrate. New York: Interscience.
5. Southern, E. M. (1975) Detection of specific sequences among DNA fragments separated by gel electrophoresis. J. Mol. Biol. 98:503–517.
6. Goldberg, D. A. (1980). Isolation and partial characterization of the *Drosophila* alcohol dehydrogenase gene. Proc. Natl. Acad. Sci. 77:5794–5799.
7. Towbin, H., Staehelin, T., and Gordon, J. (1979) Electrophoretic transfer of proteins from polyacrylamide gels to nitrocellulose sheets–procedure and some applications. Proc. Natl. Acad. Sci. 76:4350–5354.
8. Zuk, R. and Litman, D. J. (1984) Immunochromatographic assay with support having bound "MIP" and second enzyme. US Patent 4,435,404.
9. Campbell, R. L., Wagner, D., and O'Connell, J. P. (1987) Solid phase assay with visual readout. US Patent 4,703,017.
10. Rosenstein, R. W. and Bloomster, T. G. (1989) Solid phase assay employing capillary flow. US Patent 4,855,240.
11. Mansfield, M. A. (2005) The use of nitrocellulose membranes in lateral-flow assays. In: Wong, R. C. and Tse, H. Y., eds. Forensic Science and Medicine: Drugs of Abuse: Body Fluid Testing. Totowa, NJ: Humana Press, pp. 71–85.
12. Weiss, A. (1999) Concurrent engineering for lateral-flow diagnostics. IVD Technology 5(7):48–57.
13. Chandler, J., Gurmin, T., and Robinson, N. (2000) The place of gold in rapid tests. IVD Technology 6(2):37–49.

Chapter 7
FUSION 5: A New Platform For Lateral Flow Immunoassay Tests

Kevin Jones

7.1 Introduction

In recent years, there has been an explosion in the use of rapid assays. In particular, the ease of use of lateral flow immunoassays has revolutionized the area of personal healthcare. Lateral flow immunoassay tests have become the format of choice for many testing situations such as pregnancy, ovulation, drug use (both therapeutic and illicit), infectious disease, environmental health, and even in the bio-defense and forensics markets.

The key feature of a lateral flow test is that it is very simple for the end user. Normally, the user simply has to apply the sample and the test will run within a matter of minutes. In reality, the lateral flow test is a much more complicated medical device than many users appreciate. Within the plastic housing there can be five different membrane or fibrous components assembled onto a plastic backing. There are also at least two antibodies, an antibody conjugate, buffers, and blocking agents as well (see Chapters 4 and 8). Choosing the compatible solid phase materials and applying the biological components in a way that results in a sensitive and stable assay can be a major challenge. For this reason, a new material termed FUSION 5 is designed to overcome many of these obstacles.

7.2 Comparison of FUSION 5 with a Traditional Lateral Flow Strip

FUSION 5 can be used as a replacement for all the solid phase materials currently used in a lateral flow assay. It is a product that allows an entire lateral flow test to be built upon a single material, simplifying the manufacturing process and providing a significant cost-saving to manufacturers. FUSION 5

K. Jones (✉)
Whatman Inc., Florham Park, NJ
e-mail: Michael_mansfield@millipore.com

R.C. Wong, H.Y. Tse (eds.), *Lateral Flow Immunoassay*,
DOI 10.1007/978-1-59745-240-3_7, © Humana Press, New York, NY 2009

also overcomes some of the limitations seen with traditional materials. It is very fast wicking, so test times are reduced. It is naturally hydrophilic, so it does not require extensive blocking or careful storage conditions. Furthermore, it does not substantially age on storage. The target in the initial phase of FUSION 5 development was to produce a single solid phase material that would allow a typical assay where up to 100 μl of whole blood was applied and for the test to be completed within a 4-min period without any significant loss of sensitivity or specificity as compared to existing tests. In comparison, current whole blood assays based upon nitrocellulose can take between 15 and 20 min for the assay to finish (Fig. 7.1).

In terms of structure, FUSION 5 is a glass fiber–based material that contains a plastic binder to increase mechanical strength and to maintain the physical properties such as hydrophilicity and wicking speed during storage. FUSION 5 is unique as it can work as any part of a lateral flow system. It can function as a blood separator, a sample pad, a conjugate release material, a reaction substrate or an absorbent pad. It can also function in any combination or all of those parts at the same time.

The most common problems seen with the manufacture of lateral flow devices today are caused by the hidden complexity of the device. As there are so many components present, problems can be caused by material compatibility issues, contact problems, or imperfect material characteristics.

7.2.1 Material Compatibility Issues

In a typical lateral flow system, there are many different types of materials that can be used. These include cast membranes (e.g., nitrocellulose,

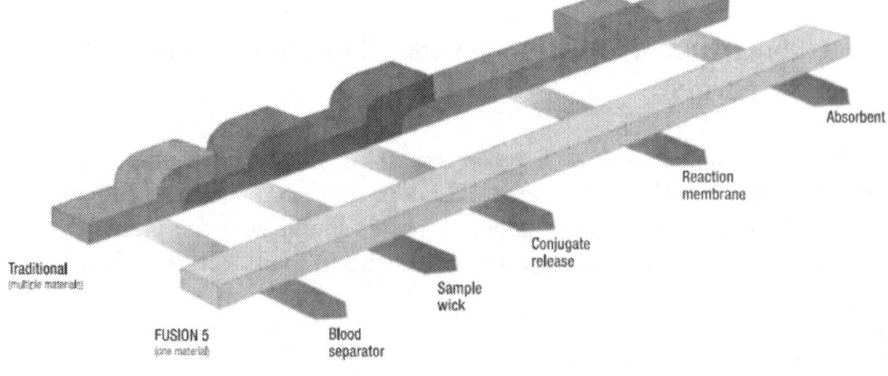

Fig. 7.1 Comparison of a traditional lateral flow test with one built upon FUSION 5

polyethersulfone), fibrous filters (e.g., cellulose, glass fiber, rayon, polyester), filters manufactured from fused particles (e.g., polyethylene), and even some woven polymers. Each of these materials has different surface properties, wicking characteristics, hardness, and shelf-life and storage conditions. They also require different handling and processing during manufacturing. Choosing the best combination of materials can, therefore, be very difficult and time consuming. There can be many unwelcome surprises as the composite test system ages. Many tests have been developed where poor material compatibility of the solid phase materials has resulted in uneven sample flow through the system, leading to poor line quality or erroneous results.

7.2.2 Contact Problems

The traditional test strip is made up of all the different components laid on top of each other with a well-defined overlap. There has to be correct pressure between each component at the contact points to ensure good transport as the sample flows along the test device. If the applied pressure is too low, the fluid will not flow from one material to the next. If there is too much pressure the materials can be crushed, which will impede flow. So at each contact point along the strip, there has to be a pressure bar designed into the housing to ensure the perfect contact that is required. This can cause significant problems as all of the materials used have different thicknesses and are of different hardness. Therefore, the amount of pressure required for the contact will vary between different materials. If the materials change, a different pressure will be required. The actual pressure (controlled by the size of the pressure bar in the housing) will hence have to be optimized for each set of materials, and it can be very difficult to use the same housing for a range of assays.

7.2.3 Imperfect Material Characteristics

The materials that are normally used in lateral flow strips are not originally designed for this purpose. Cellulose or glass fibers were developed for liquid filtration. Rayon and non-woven polyesters were developed for the clothing or furniture business, and the large-pore nitrocellulose membranes were originally developed for air filtration. In the end, developers have made use of these special properties and built them into a rapid assay. However, while the materials may be adequate for most applications, their performance is not perfect. A good example is the use of nitrocellulose.

Nitrocellulose is so far the best cast membrane for lateral flow assays (see Chapter 6). It has performed well for 20 years, and no other membranes have been able to match the signal-to-noise ratio seen with nitrocellulose membranes. On the other hand, nitrocellulose does have some problematic drawbacks when used in lateral flow assays:

- Nitrocellulose is naturally hydrophobic. The only reason that liquids will flow through the membrane is because of the addition of hydrophilic materials such as surfactants or polymers that encourage fluid flow. These hydrophilic materials can interfere with binding reactions at the capture line and/or interact with other components in the system. As the composition and concentration of these hydrophilic materials are proprietary to the membrane manufacturer, it can be hard for developers to adequately predict their effects on the test system.
- Nitrocellulose also ages. The wicking rate and protein-binding characteristics of the membrane change with time. Aging of the membrane will have a significant effect upon long-term assay performance.
- Nitrocellulose binds hydrophobic materials. One of the main reasons for using nitrocellulose is that it binds hydrophobic materials, like proteins, irreversibly. However, the ability to bind hydrophobic materials can also give rise to other problems. Nitrocellulose will bind the specific proteins for the lateral flow assays, forming the capture or test line. But it also has the potential to bind other proteins or the conjugates used in the rapid assay as well. To overcome this problem, surfactants or polymers are added to the test system to stop non-specific binding of hydrophobic materials. As pointed out above, the presence of surfactants and polymers can interfere with the test assays, particularly the immunogenic interactions that are needed for proper functions of the assay. Presence of high levels of hydrophilic blocking materials can result in false positive or negative results.

Therefore, a material like nitrocellulose that is used by practically every manufacturer of lateral flow immunoassays does have some issues with use. Other materials that are used in lateral flow assays have similar undesirable properties. Even though they do perform the desired role well, they have to be modified in some way to ensure that the side effects are controlled and do not interfere with the way the assay works. FUSION 5 overcomes many of these problems without requiring any significant changes in the technology. A direct comparison of FUSION 5 with traditional techniques and materials are shown in Tables 7.1 and 7.2. Because FUSION 5 was designed for use in lateral flow immunoassays, and is very tightly controlled during manufacturing, there are less unexpected problems when using it. FUSION 5 is non-protein and non-conjugate binding. It is naturally hydrophobic, and it does not require blocking agents. Furthermore, it does not age. As a result, FUSION 5 does not exhibit many of the problems seen with the traditional materials used for lateral flow immunoassays.

Table 7.1 Comparison of traditional test strip technology with FUSION 5

Traditional strips	FUSION 5
The test strip is made from multiple materials	The test strip is made from one material
Each functional component is treated in a separate manufacturing step	Only one manufacturing step
Requires assembly of the finished components into a strip	No assembly required
Requires the correct contact pressure to make the sample flow	No contact pressure points required

Table 7.2 Comparison of FUSION 5 and nitrocellulose

Nitrocellulose	FUSION 5
Protein binding under normal test conditions	Non-protein binding under normal test conditions
Conjugate binding under normal test conditions	Non-conjugate binding under normal test conditions
Naturally hydrophobic	Naturally hydrophilic
Requires blocking	No blocking step required
Requires surfactant in the assay	No surfactant needed*
A significant aging effect is observed	No significant aging observed

*Unless surfactants are required to stop non-specific binding between assay components.

7.3 Performance of Fusion 5 in Lateral Flow Immunoassay Systems

However good a material may sound in theory, the real proof lies in how the material actually performs in an assay system. FUSION 5 is a glass fiber–based material and there is nothing novel in the use of FUSION 5 as an absorbent or wicking material as glass fibers have been used in these roles for many years. There are, however, three key areas where FUSION 5 performance will have a substantial effect upon the way the test system works.

7.3.1 FUSION 5 as a Blood Separator

It has been known for many years that glass fiber can work effectively as a blood separator. Red cells are attracted to the surface of the glass fiber and become trapped within the depth of the matrix. When a sample of fresh blood is applied to FUSION 5, it was found that the composition of the serum produced was similar to that of the serum separated by the conventional centrifugation method (Fig. 7.2). The main feature of FUSION 5 is that it can produce greater than 21 µl of serum from a 40 µl drop of blood within 1 min of the blood drop being applied. This is almost 93% efficiency based upon the volume of blood obtained by centrifugation. The proportions of key analytes are unchanged by filtration through FUSION 5. Many of the typical targets for immunoassays

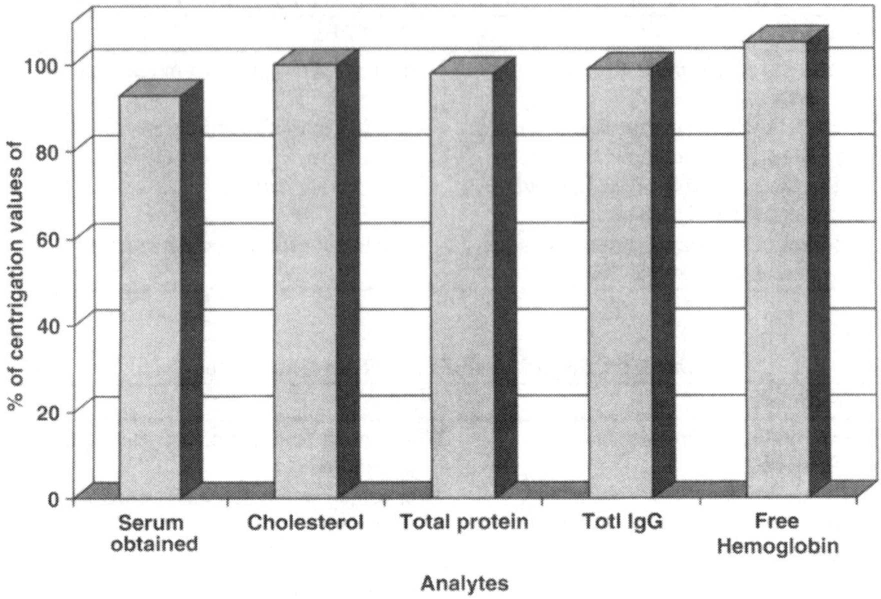

Fig. 7.2 Comparison of serum components produced by FUSION 5 and by centrifugation (numbers expressed as a percentage of the centrifugation values)

(e.g., cholesterol, total protein, and IgG) have essentially the same properties as compared to centrifuged blood. These results indicate that assay performance by FUSION 5 can be extrapolated across test platforms.

7.3.2 FUSION 5 as a Conjugate Release Pad

The conjugate release pad is, in theory, a simple part of the assay. Its function is to allow a colored particle to dry onto a solid phase without being damaged, and to release it when a sample is applied so that a colored line can appear on the reaction membrane. In practice, however, proper usage of a conjugate pad is one of the most difficult aspects of producing lateral flow immunoassay tests. The reasons for this difficulty are not necessarily caused by the conjugate pad itself, but rather by the nature of colloidal particles used in the test system (see Chapter 5). In solution, colloidal conjugates are very stable. Almost all colloids can be stable in solutions in excess of a year, and not uncommonly for over 5 years. Colloidal gold and latex particles are stabilized in solution due to charge repulsion. However, the colloids used in a lateral flow immunoassay are not in solution form. They are in a suspension. During the drying process, the particles are forced close together with the potential to touch each other. When the particles touch, they aggregate permanently. Thus, the drying step is a major cause of performance problems. Once aggregates are formed in the conjugate pad, it is very difficult to separate them again.

All the common forms of colored colloidal particles used in lateral flow immunoassays have similar physical characteristics. They are all negatively charged, hydrophobic particles, which have been typically coated with an antibody to allow interaction with the target analyte. FUSION 5 works well as a conjugate release pad as it is negatively charged and also hydrophilic. As a result, there is no strong attractive force between FUSION 5 and the colloidal particles used, making release of the conjugates from the surface of FUSION 5 as favorable as possible.

FUSION 5 has been tested with both gold and latex antibody conjugates, and has given good performance with each type. The performance of FUSION 5 is seen to be similar to other types of conjugate pad commonly used (Fig. 7.3). To minimize problems when using FUSION 5 as a conjugate pad, the following procedures are recommended:

- Include a low level of hydrophilic polymer or surfactant to reduce hydrophobic interaction between the particles.
- Use very low ionic strength solutions. As the conjugate pad dries, the effective salt concentration increases and this causes particle aggregation.
- Minimize the concentration of proteins used as they can cause pad hydrophobicity upon drying.
- Maintain the pH of the conjugates above the pI of the proteins to ensure that all components stay negatively charged. If both the colloidal particles and FUSION 5 are negatively charged but the proteins are positively charged, the particles will stick to each other and to the surface of the FUSION 5. Table 7.3 shows the typical buffer systems recommended for use with FUSION 5.

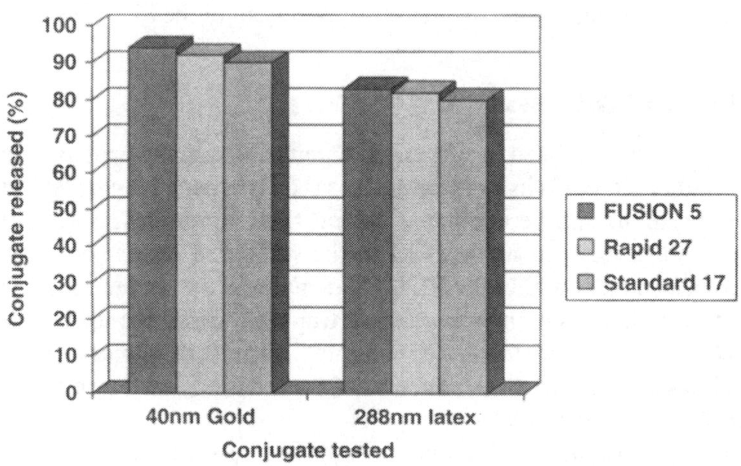

Fig. 7.3 Release of conjugates from FUSION 5 verses other common conjugate pads

Table 7.3 Typical buffers for use with FUSION 5 as a conjugate release

	40 nm Gold conjugate	300 nm Latex conjugate
Typical buffer	50 mM phosphate pH 7.2 + 1% BSA + 1% Tween 20 + 0.5% PVA	50 mM phosphate pH 7.2 + 1% BSA + 1% Triton X-100 + 1% PVA + 0.3% PVP
Typical performance	>94% conjugate release after 45 seconds	>83% conjugate release after 45 seconds

Finally, due to the fact that FUSION 5 is recommended to work with a single manufacture step, traditional dip coating applications are not recommended. Spraying the conjugate onto the material from an airbrush or striping the conjugate through a wide diameter tube gives excellent results. The application of the conjugate to the surface of the material results in a very rapid and complete release. In comparison, dip coating saturates the entire pad with conjugates, resulting in a slower and less complete release.

7.3.3 FUSION 5 as a Reaction Membrane

From the above discussion, it can be seen that FUSION 5 works well as a blood separator, a sample pad, or a conjugate release pad. However, the key property for getting a good-quality test line on a lateral flow assay when using FUSION 5 is how well the material functions as a reaction membrane. The relevant question for many users is how strong and sharp is the capture line as applied. Under normal conditions FUSION 5 is non-protein binding. To make proteins bind to FUSION 5, it is necessary to find alternative routes to allow formation of the capture line. Two ways have been used to ensure that proteins can be bound to the surface of FUSION 5. These are use of a carrier bead and applying proteins at low pH. Both techniques work and are suited to different applications of the FUSION 5 material.

7.3.3.1 Use of a Carrier Bead

The concept of using a bead in a large-pore material to make proteins bind was first reported in the 1970s by systems such as the Hybritech Icon or Abbot ABx. The same principle can be applied to lateral flow immunoassay tests. In this case, the capture reagents are applied to the surface of carrier beads and the beads become trapped within the FUSION 5 material due to their size. The size of the bead is critical as it ensures efficient trapping within the matrix and also provides the surface where the test lines appear. Therefore, choice of the correct bead size is a key factor in the production of a sensitive assay. The effect of bead diameter on the assay is shown in Fig. 7.4.

Technically, if a small bead is used, then the line becomes broader as the latex beads migrate a small distance through the matrix before they become trapped. However, the line produced is relatively strong. As shown in Fig. 7.4, a 2 μm

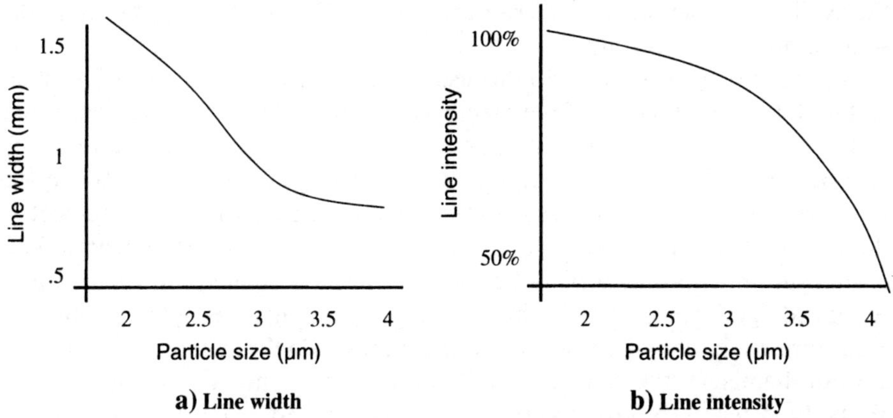

Fig. 7.4 The effect of carrier bead diameter on the performance of FUSION 5 line width

latex bead will give a line width of approximately 1.5 mm, but the relative intensity will be 100%. On the other hand, if a large bead is used, the line is narrower as the particles are trapped in where they are applied and the intensity is seen to drop. This reduction in intensity is due to the available surface area on the beads. Large beads have a lower "relative" surface area in a given volume than small beads. Thus, a 4-μm latex bead will have a line width of 0.7 mm, but the relative intensity is only about 50% of the small bead. The best compromise is seen with beads of approximately 2.3 μm. In this case, there is a sharp line (1 mm) that has high signal intensity (95% of maximum intensity). The reasons for this effect are explained below.

The line intensity seen on FUSION 5 can be compared with that seen on slower-wicking nitrocellulose membranes. The line intensity is normally proportional to the surface area available for reaction. So a material with a large surface area will allow the application of more capture reagents, leading to the capture of more antigen and detection reagent than a material with lower surface area. In nitrocellulose membranes, the wicking rate is related to surface area. As the wicking rate increases, the surface area decreases. However, in FUSION 5, the surface to which the capture reagent is bound is often a bead. When using beads, the available surface area is related to the surface area of the bead and the number of beads present in a given volume of capture line. The choice of bead diameter therefore becomes a key factor in optimizing the capture line intensity on FUSION 5. The smaller the bead diameter, the greater the "relative" surface area is for a given volume. Large beads have a lower relative surface area than small beads, and the potential sensitivity will therefore decrease. Using a 2-μm latex bead with FUSION 5 gives approximately five times the surface area that is seen on a nitrocellulose with a nominal pore size of 8 μm. So there can be five times the amount of capture reagent immobilized if required for the assay. If the diameter of the capture bead is increased to 4 μm, the available surface area at the capture line of FUSION 5 would be about the same as an 8-µm nitrocellulose.

An additional advantage with the use of carrier beads is that, in a traditional material, the void volume of the membrane is large. Basically, this means that there are areas of the membrane where the distance between each wall of the membrane is relatively large, thus producing a large open space between the walls. This is significant as the size of the membrane pore is large with respect to the size of the capture reagent or the conjugates. A typical nitrocellulose membrane with a wicking rate of 80 seconds has a nominal pore size of 8 µm. A typical IgG molecule is 8 nm, and a gold conjugate is typically 40 nm. Therefore, as the conjugate flows through the pores in the membrane (assuming a 8 µm diameter), the chance of a 40-nm conjugate interacting with a 8-nm antibody is relatively low. Interaction will occur only when the conjugate is close to the walls of the pore. When beads are used in a FUSION 5 membrane, the voids are filled with beads. The distance from the center of the void surface where the capture reagents are dried is therefore significantly reduced, and the chance of interaction increases. The use of smaller beads gives better packing than when using larger beads, so the voids are smallest when using small beads.

The nature of the carrier bead used is not really significant. Provided that the carrier bead allows the stable addition of protein to the surface, and allows the sample to approach the surface even after prolonged storage, then any kind of beads can be used. Currently, most of the work has been carried out using polystyrene beads with physical adsorption. However, the use of silica, cellulose, dextrose, or almost any other material would be acceptable as well. Normally, the beads would be the same color as the membrane so that there is no visible color tint at the test line before the sample is run. However, one potential benefit can be obtained by using a capture bead that has a faint, light color (e.g., a pale yellow bead). This will allow the location of the test line to be identified before the test is run, making the reading of multi-analyte tests more straightforward. This would also help with QC as a visible marker that the bead materials have indeed been striped on the membrane during manufacturing.

It is possible to use standard application systems to spray or stripe the beads onto the FUSION 5 membrane. For example, beads of up to 6 µm can be applied through a spray system such as the BioDot BioJet ® (Irvine, CA), and through a contact applicator such as the BioDot Frontline or the systems from Imagene (Seoul, Korea). Therefore, the existing systems used for putting protein or conjugate lines onto lateral flow membranes can be used without any need for modification.

7.3.3.2 Application of Proteins at Low pH

FUSION 5 has a permanent negative charge. The proteins that are applied onto FUSION 5 can change their charge depending on the pH of the buffer used. It is therefore possible to make most proteins positively charged simply by changing the buffer in which they are stored. Spraying the proteins onto

FUSION 5 at low pH will result in adhesion of the proteins to the surface. Drying by normal methods will then result in permanent attachment of the proteins to the matrix. Typically, the pH of the system needs to be reduced to approximately 3. However, this also depends on the nature of the proteins being applied. The pH must be low enough to ensure overall positive charge, without adversely affecting the immunogenicity of the proteins or causing their aggregation.

7.3.4 When to Use Each Type of Application Techniques

In principle, the use of carrier beads to fill the membrane pores will result in increased sensitivity due to an increase in the available surface area and the filling of the voids in the material. These procedures, however, do add extra time and cost to the production process. Spraying the proteins directly onto the membrane is much more straightforward, but the drawback is that the surface area is much lower. Therefore, if sensitivity is of high concern, it is recommended to use carrier beads. If manufacturing cost is more important, direct application is a much cheaper option.

7.4 Trouble Shooting FUSION 5

FUSION 5 has been shown to work well for use in all aspects of a lateral flow system as an individual component. However, some issues may arise when it is used as a complete system. The following are some common problems that are relatively easy to rectify.

- Sample does not wick through the test: As FUSION 5 is hydrophilic, this problem is caused by the introduction of some hydrophobic components. Typically, this is through the use of high levels of proteins being applied for blocking procedures. If blocking is required, the use of polymers or surfactants (e.g., polyvinylalcohol, poly vinylpyrrolidone, Tween 29, or Triton X-100) is preferable. Hydrophobicity can also be caused by the transfer of plasticizers or mould-release agents from plastic bags or molded housings.
- Red cells passing through the FUSION 5: If a red color is seen passing through the FUSION 5 material, this can be due to either red cells passing through the matrix or hemolysis. The most common reason for red cells to migrate through the matrix is due to the excessive pressure being applied to the FUSION 5. The pressure over the blood separation area should therefore be kept low. It has also been shown that red cells age on storage. To reduce the risk of hemolysis, male blood should be used for only 24 hours after collection whereas female blood can be used for up to 3 days. Inclusion of a small amount of sodium acetate or ascorbic acid in the sample collection area

can reduce the level of hemolysis seen with aged blood. Hemolysis is not commonly seen when using fresh blood.

- Poor conjugate release: The reasons for poor conjugate release are several. This may be caused either by the use of a conjugate that is prone to aggregation or the use of high salt or low pH buffers during conjugate application.
- Low capture line intensity: This can be caused by incorrect application of the capture line or by the test running too quickly. Make sure bead sizes are not too large. Beads between 2 and 4 μm in diameter can be used effectively with FUSION 5. However, it must be remembered that as bead size increases the line intensity decreases (see Fig. 7.4 above). Low capture line intensity can also be due to only low concentrations of capture reagents that are immobilized to the beads. In some covalent linking techniques, there is only a low level of protein bound to the surface of the bead. This will therefore provide only a low chance of interaction with the analyte and conjugate when the assay is run. Finally, FUSION 5 is very fast wicking. The wicking rate seen is much higher than with nitrocellulose membranes. As the wicking rate is so fast, the time for interaction at the capture line is very short. For some assays, it is necessary to slow down the wicking rate to enable a greater chance for binding at the capture zone. The addition of long-chain polymeric materials (e.g., hydroxylpropylmethylcellulose; high molecular weight PVA or PVP) will increase the viscosity of the liquids, and this will have a significant impact upon the line intensity that can be observed.

7.5 Cost Comparison of FUSION 5 in Lateral Flow Assays

Besides having performance advantages over traditional strips, FUSION 5 can also have a significant impact upon the costs of lateral flow tests production. The key advantages of FUSION 5 over traditional materials are simplicity and speed. In traditional assays, there are a number of different materials used to produce the test strips. With FUSION 5, there is one single material used. Basically, the use of FUSION 5 can reduce the time associated with manufacturing lateral flow strips. This time reduction obviously has a direct relationship to the final costs of production.

7.5.1 Production of Traditional Test Strips

The first step in the manufacturing of a traditional assay test strip is coating of the conjugate pad. The original method for doing this was to dip coat (or impregnate) the conjugate pad with conjugated materials and then dry (see Chapter 8). Drying of the conjugate pad is a very time-consuming process, as there is a great deal of water to drive off. This can possibly be the most time-consuming part of the whole manufacture process, and is one of the most difficult due to the low mechanical strength of many conjugate pads when wet. In an attempt to reduce production time, some companies have now adapted production techniques by spraying the conjugate materials onto the conjugate pad using traditional application systems. This has been a success in terms of reducing manufacturing time, and had the unexpected benefit of reducing the variation in the amount of conjugates released from the pad. Typically, these systems can run at a speed of 0.5 m/minute. The major limitation is the drying efficiency for the conjugate pad.

Following drying of the conjugate pad, the next process is to apply the test and control lines. This is performed on the application systems made by companies such as BioDot (Irvine, CA), Imagene (Seoul, Korea), or Kinematic (Sonora, CA). In these systems, the reagents are sprayed or striped onto the membranes, typically running at a speed of up to 1 m/minute.

The final step of production is strip assembly or lamination. These procedures are relatively time-consuming and are prone to operator errors. Strips of various pad components need to be stuck onto an adhesive vinyl backing, correctly aligned with appropriate amount of overlaps to facilitate the flow of sample along the strip. Typically, this step is performed manually by placing strips of the material in a holder, and sticking the components onto an adhesive vinyl backing card. A skilled operator can perform three to four assembly operations per minute. After assembly is completed, the cards can be cut into strips for placement in the plastic housings.

7.5.2 Production of FUSION 5 Strips

In contrast to the complexity of manufacturing standard test strips, FUSION 5 is very simple to manufacture. Bottles of conjugates and capture reagents are placed onto the spraying machine (e.g., BioDot, Imagene or Kinematic) and a roll of FUSION 5 is run through the machine. The speed of the system needs to be optimized, but typical speeds in excess of 0.5 m/minute are used. Again, the major limitation is the drying required by the conjugate area. More efficient drying can result in faster operating speeds. If necessary, additional spray heads can be used to apply blocking reagents. As FUSION 5 is naturally non-protein and non-conjugate binding, the requirement for additional blocking is not always necessary.

After drying, the reel of FUSION 5 can simply be cut into the strips for placement into housing directly, or laminated into plastic films to make a housing-free assay. The time-consuming manual steps of assembly are eliminated from the process, resulting in a more efficient, less labor-intensive and therefore more cost-effective product. Cost comparison of using FUSION 5 against a traditional style test is listed in Table 7.4.

Table 7.4 Cost comparison of using FUSION 5 against a traditional style test (labor cost per linear meter)

	Traditional (impregnation) test	Traditional (spray) tests	FUSION 5
Coat and dry the conjugate pad	4 minutes 10 seconds	2 minutes	2 minutes (single-step manufacture)
Strip and dry the membrane	1 minute 40 seconds	1 minute 40 seconds	
Laminate the strip	1 minute 10 seconds	1 minute 10 seconds	
Total time per lin m	7 minutes	4 minutes 50 seconds	2 minutes
Total cost per lin m*	$0.834 per lin m	$0.576 per lin m	$0.238 per lin m
Cost saving on labor using FUSION 5*	71.9%	58.6%	

*Cost calculated assuming 2007 NY State minimum rate of $7.15 per hour.

7.6 Conclusions

FUSION 5 provides the manufacturers of lateral flow immunoassays an alternative platform for development. It is fast wicking, has a low background, and produces assays at a significantly lower cost than using the traditional approaches. Speed and simplicity are the best advantages of using FUSION 5. However, designing a new test system with FUSION 5 requires different techniques and different approaches as opposed to those used for traditional assays. Developers have to be willing to learn the best way of using the material.

Appendix: Steps in Using FUSION 5 for a Whole Blood hCG Test

1. Start with a reel of FUSION 5 that is 6 cm wide and 50 m long.
2. **Test line:** At the capture zone (approximately 2.5 cm from the edge of the material), apply 2-micron latex beads conjugated to a monoclonal anti-alpha hCG antibody to the FUSION 5 matrix with 10 mM phosphate at pH 7.2.
3. **Control line:** At the control zone (approximately 2.7 cm from the edge of the material), apply 2-micron latex beads conjugated to a goat anti-Mouse IgG antibody to the FUSION 5 matrix with 10 mM phosphate at pH 7.2.
4. **Conjugate release:** At the conjugate release zone (approximately 1 cm from the edge of the material), apply a 150 nm blue-dyed latex colloid conjugated to a monoclonal anti-beta hCG antibody to the FUSION 5 matrix with 10 mM phosphate at pH 7.2 + 0.5% Tween 20 + 0.5% PVA + 0.2% BSA.

After application of the lines (that may be done simultaneously), dry the striped FUSION 5 material for at least 3 hours at 37°C.

Cut the FUSION 5 into 5 mm-wide strips, and place the materials in a plastic housing. Apply 100 µl of whole blood to the test strip. As the whole blood sample wicks up the strip, the cellular component of the blood will be captured by the FUSION 5, allowing the cellular component to flow without the red cells. The acellular component will reach the conjugate release zone and resuspend the blue latex conjugate. Any hCG present will interact with the anti-beta hCG antibodies present on the conjugate. The sample and resuspended conjugate will continue to flow to the capture line, where any hCG present will attach to the anti-alpha hCG antibody present on the capture line, forming a sandwich resulting in the dyed latex being retained at the capture line. A positive result (indicated by the appearance of a blue line at the capture line) will indicate the presence of elevated levels of hCG (a hormone associated with pregnancy in humans). At the control line, the anti-mouse IgG will interact with the monoclonal antibody on the conjugate. The control line will turn blue whether there are elevated levels of hCG present or not.

Note: To provide additional mechanical strength, laminate the FUSION 5 matrix to a vinyl backing card coated with a pressure adhesive. These would be the same cards as those used for the assembly of traditional lateral flow strips.

Chapter 8
Manufacturing the Next Generation of Highly Sensitive and Reproducible Lateral Flow Immunoassay

Thomas C. Tisone and Brendan O'Farrell

8.1 Introduction

The lateral flow immunoassay (LFIA) format that emerged in the 1980s provided the industry with a relatively standardized, open platform that could accommodate a wide range of analytes for point-of-need (PON) testing. Figure 8.1 shows a typical configuration for a LFIA strip. LFIA has been applied in application areas as diverse as human clinical diagnostics, veterinary diagnostics, plant health, environmental health and safety, food safety, and biowarfare, just to name a few [1]. To date, the application of LFIA has largely been limited to threshold or yes/no formats due to the high levels of variation exhibited with the current designs and manufacturing processes. Despite this limitation, it is estimated that over 100 different tests are currently available in the market. The ability to adapt the LFIA to a quantitative format is predicted to increase the number of marketable tests significantly.

In its nascent stages, the intellectual property and large-scale manufacture of point-of-care (POC) immunoassays were dominated by a few large companies with significant infrastructure. Over the past 20 years, however, the majority of the assays developed and much of the resulting market development has been driven by much smaller entities. The LFIA market has evolved considerably in the past decade, with consolidation of companies, identification of new market opportunities, increases in the number of strips produced worldwide, and the emergence of new technology to address evolving market needs.

The performance requirement for PON assays vary considerably from market segment to market segment. In the most general form, the performance of LFIAs has always been challenged in two major areas: reproducibility and sensitivity. The lack of strip-to-strip reproducibility has dogged the reputation of this assay format and prevented its adoption and application in many quantitative settings. Much of this lack of reproducibility may come from the materials used in the

T.C. Tisone (✉)
BioDot, Inc., Irvine, CA
e-mail: tom.tisone@biodot.com

R.C. Wong, H.Y. Tse (eds.), *Lateral Flow Immunoassay*,
DOI 10.1007/978-1-59745-240-3_8, © Humana Press, New York, NY 2009

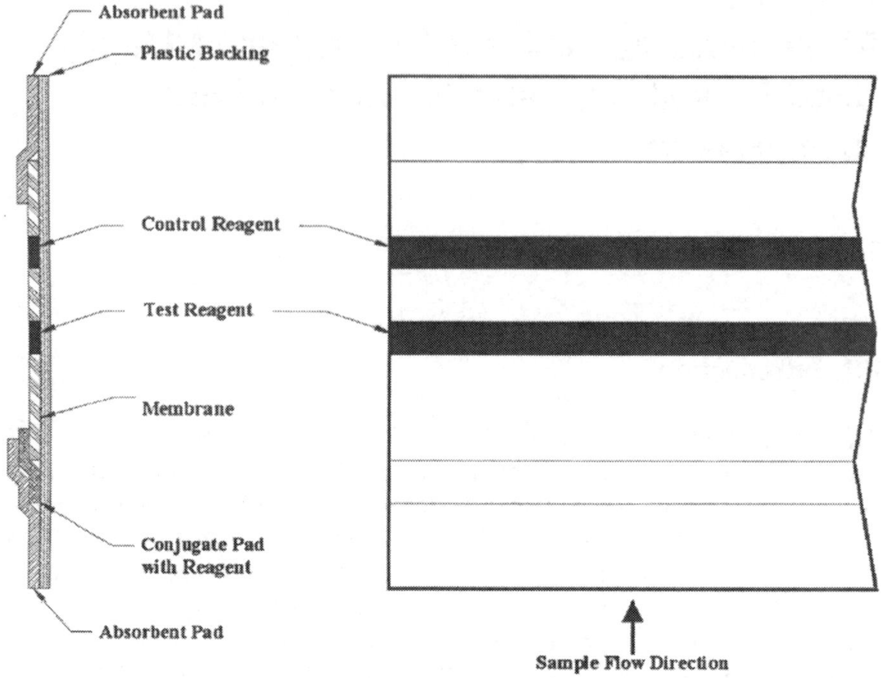

Fig. 8.1 Structure of a typical LFIA assay

assays. However, the manufacturing tools and processes used also account for a considerable share of this inter-assay variation [1]. The other performance element that is being driven to higher levels by market needs is sensitivity. This is a general trend in the diagnostic marketplace for all assay formats. In the context of considering manufacturing processes and tools, it is interesting to note that, rather than being a function of the manufacturing processes, the requirement for increased sensitivity is imposing new standards on the manufacturing processes. Several aspects have contributed to the drive for change:

- *Continued improvement in product quality*: The major issue is to continue to improve manufacturing coefficients of variations (CVs) with improvements in processes, materials, and labels used for assembly of the LFIA [1, 3].
- *Development of new production tools and methods*: These need to support both improved manufacturing CVs for current LFIA formats as well as newly evolving LFIA platforms. Processes need to be verifiable and quantitative, relevant to both reproducibility and regulatory considerations [1–3].
- *Ability to produce higher sensitivity and quantitative test formats*: This includes new labeling systems, new materials, and new processes and tools to support these developments. New labeling systems include new conjugate formats and readers [1, 3], such as the magnetic particle and magnetic reader system of

Magna Bioscience [4] and novel fluorescence assays such as those reported by Diagnostic Consulting Network [5] and Amic [6]. ChemBio has also introduced a new lateral flow format based on a double-flow path [7]. These types of devices are based on design modifications to the classical LFIA format and present new sets of assembly and process requirements and challenges.

The consolidation of manufacturing in the diagnostics industry, the expansion of the technology into non-traditional markets with potentially high volume requirements, and the entry of large companies from other industries that have traditionally utilized highly controlled production processes are all factors that have impacted both the anticipated throughput requirements of the next-generation assays and applications and the means by which high volumes of these products must be produced. Two aspects must be considered:

- *Transfer from Research and Development to Manufacture*: Time to market is an important consideration in today's international markets. This means that the time from recognition of a market opportunity to the ability to ship product needs to be as short as possible. One practical means of minimizing the time for manufacturing transfer, scaling up and revalidation of high-volume production processes, is to ensure that the development of new products is performed on production-rated equipment and processes.
- *Scaling processes and equipment to volumes in excess of 100 million per year*: Markets in some areas of health care and food are expected to reach sizes of billions of tests per year. Theranostic markets, where a diagnostic test is paired with the delivery of pharmaceuticals either as a test of susceptibility or efficacy, are examples of a potential area that is expected to generate demands in the billions of units per year range. These types of market will require manufacturing systems that can run 24 hours per day, 7 days per week to meet the production requirements.

This chapter will first provide an overview of the commonly used manufacturing process technologies for LFIAs of traditional design. We will then introduce improved manufacturing processes that address high-throughput manufacturing requirements and tighter controls of product quality in traditional LFIA designs. Finally, we will discuss the evolution of new product requirements and test designs and their effects on the manufacturing processes and related tools.

8.2 Traditional Manufacturing Processes of LFIA

During the 1990s, the LFIA format was rapidly adapted by a large number of entrepreneurial companies around the world, responding to numerous market opportunities. Even though the core intellectual property associated with the technology was concentrated in the hands of a few major companies [for example, 8, 9], development and manufacturing of LFIAs were rapidly picked up by dozens of smaller companies. As a result, many of the manufacturing processes were developed in a relatively uncontrolled manner. The major attraction of the LFIA format for diagnostic test suppliers is the simplicity of design, which allows them to develop and manufacture easy-to-use tests with relatively high quality and very low cost.

One of the major issues inherent in the LFIA format lies in the materials used, which historically have been designed for other applications. What has evolved is a process of empirical selection of materials and pretreatments on a purely functional basis during development. This process generally results in acceptable performance for non-quantitative test formats. The basic issues of current material selection, with respect to the goals of improved CVs and quantization, are the non-uniformity of the materials in terms of dimensions, density, and chemical treatments.

8.2.1 Basic Manufacturing Processes for LFIA's

The basic processing steps for a cassette-based LFIA are shown in Fig. 8.2. These processes are outlined in the following sections.

8.2.1.1 Dispensing

The addition of complex fluids containing a variety of compounds, including proteins, surfactants, and polymers, in a controlled, reproducible way to the components of a lateral flow device is among the most technically challenging

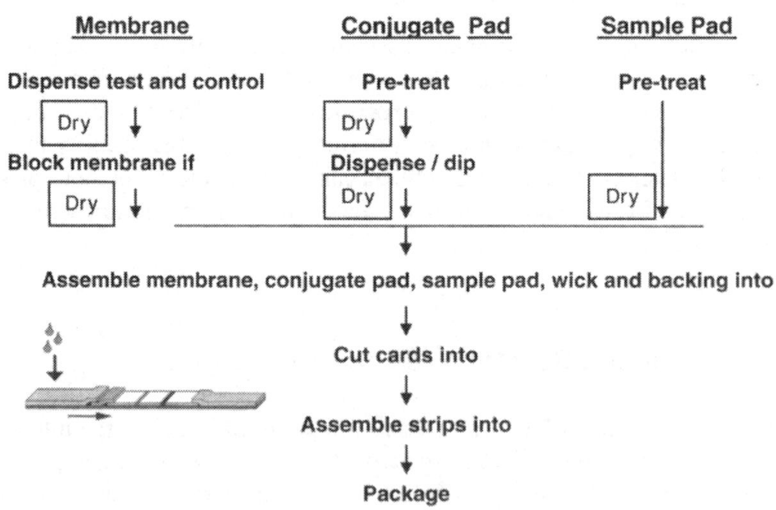

Fig. 8.2 Schematic of the manufacturing processes for LFIAs

and under-appreciated process in the production of LFIA tests. Two methods of reagent application to materials are used. For the addition of bulk volumes of fluids, such as membrane-blocking reagents, sample pad, and conjugate pad treatments, impregnation of the materials is performed by dipping into fluid tanks followed by blotting and drying. The process of impregnation will be discussed later in this chapter. For the addition of quantitative volumes of fluids in a controlled manner, such as test and control line dispensing and conjugate deposition, highly accurate and reproducible processes needed to be developed. The deposition of test and control lines on the membrane are performed using a variety of dispensing methods, which in practice have been variably quantitative and scalable on a manufacturing basis. Examples of dispenser types that have been used include off-the-shelf pneumatically driven artists' airbrushes, pneumatically driven solenoids, and positive displacement pumps with contact and near-contact tips. Previous methods performed with modified continuous ink jet printers proved to be both difficult to use and very expensive. The dispense heads were large in dimension and connected to large-diameter umbilical systems. The airbrush and solenoid technology represent relatively inexpensive non-contact formats, but were not quantitative and can be unstable due to bubble formation from dissolved air in the reagents. The syringe format uses a quantitative, positive displacement system for fluid delivery from a nozzle, but the nozzle needs to be in near contact with the membrane or material surface to get good quality, quantitative lines. This turns out to be both expensive and difficult to control in a real manufacturing environment.

What have emerged as broadly applied technologies in the industry today are three methods of dispensing quantitative volumes of reagents onto material surfaces.

i) *Non-Contact, Pump-Driven Solenoid Dispensers*: These include the BioJet Quanti [10–12] from BioDot, Inc. (Irvine, CA), which uses a drop actuator such as a solenoid hydraulically pumped with a positive displacement system (e.g., a syringe pump) as shown schematically in Fig. 8.3. This provides a non-contact ink jet dispenser where the drop volume is determined by the positive displacement. Drop-to-drop CVs can be typically in the ±5% range and short line segments comprising many drops can have CVs in the ±1–2% range. With the ink jet, dispenser lines are dispensed by overlapping drops using a linear motion system as shown in Fig. 8.4. One of the characteristics of dispensing protein solutions onto a membrane like nitrocellulose is that the protein is absorbed very quickly and essentially on impact of the drop; hence, the developed protein line in the LFIA format has a width about the size of the hemispherical drop volume as indicated in Table 8.1.

The solenoid technology can robustly dispense drop sizes down to about 15–20 nL to give line widths down in the range of about 500 μm. Continuous ink jet methods have drop sizes in the range of 1–4 nL while the drop-on-demand piezoelectric ink jet technology typically work with drop sizes in the range of 100 pL to 1 nL. The continuous ink jet technology can easily scan drop normal to the line to make either very narrow lines down into the range of 250 μm or less

Fig. 8.3 Schematic of the BioJet Quanti/Plus dispenser

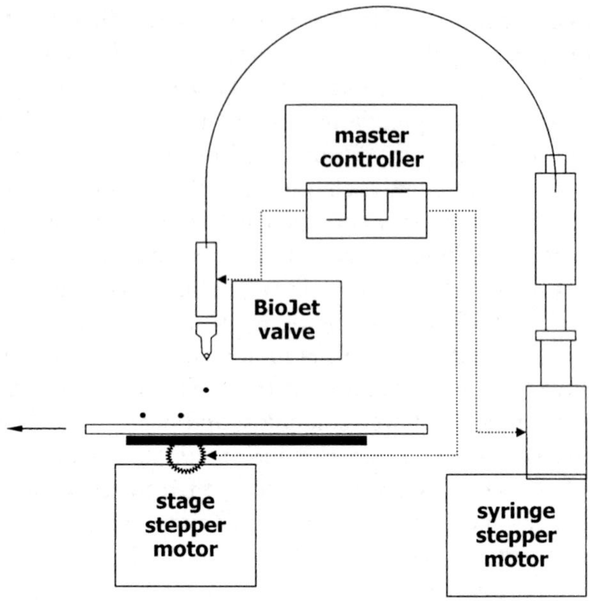

Fig. 8.4 XY Dispenser work station with BioJet Quanti and AirJet Quanti dispensers

or wider lines by overlapping many lines. The drop on demand system with a single nozzle would produce lines in the range of 100–200 μm or use arrays of dispense nozzles such as in a dot matrix printed to make wider lines. There have been many attempts to use arrays of piezoelectric drop on demand dispensers, but none have been commercially successful. Single nozzle piezoelectric dispensers are currently on the market, but are used mostly for single-drop applications such as microarrays.

Table 8.1 Drop volume vs drop spherical diameter and hemispherical diameter

Drop volume	Sphere diameter (Microns)	Hemisphere diameter (Microns)
1 pL	12.4	15.34
10	26.7	33.8
100	58	72
500	98	124
1 nL	124	156
2.08	158	199
5	212	268
10	266	337
20	336	423
50	457	577
100	575	725
500	982	1,243
1 μL	1,240	1,563

ii) *Contact Tip Dispensers*: The FrontLine gliding tip dispenser from BioDot, Inc. is one example of a contact tip dispenser. Other producers of this type of dispensing method include Isoflow™ (Imagene Technology, Inc., Hanover, NH) and Zeta (Zeta Corporation, Gunpo-City, Republic of Korea). A schematic of the FrontLine or dragging flexible tip is shown in Fig. 8.5. The dispenser consists of a syringe or other positive displacement pump with a flexible tip that now drags across the surface of the membrane. The amount of fluid pushed through a contact tip system is quantitative, as defined by the accuracy of the pump used. However, the amount of fluid absorbed per unit length of the material is not necessarily quantitative or quantifiable. This is because the absorption rate of the nitrocellulose material typically used is dependent on material, fluid and environmental factors. Material-related issues include the hydration of the membrane, the pore size of the membrane, and

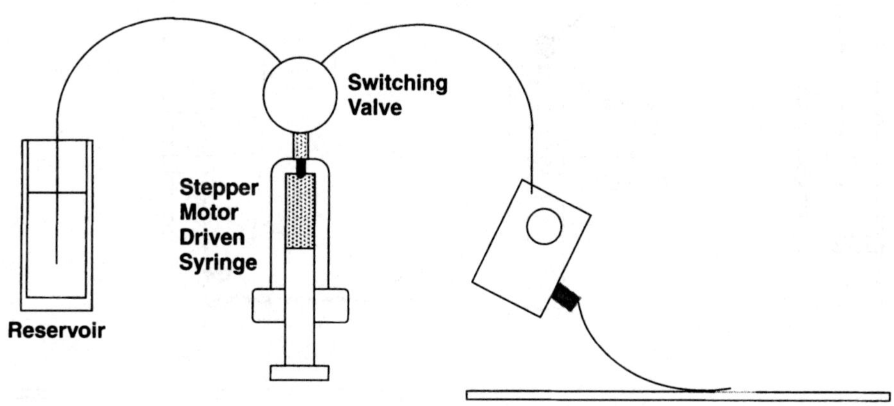

Fig. 8.5 Schematic of a dragging tip dispenser

membrane surface characteristics such as smoothness or dust. Fluid factors include protein concentration and viscosity. Environmental factors such as the ambient relative humidity of the work area also affect fluid absorption rates. These factors can result in a variation in deposited volume per unit length and developed line width. An additional drawback with contact tip dispensing is that it can perturb the membrane surface, which in turn can contribute to LFIA CVs issues [3] and introduce false positives into assays.

iii) *Quantitative Airbrush-Type Dispensers*: BioDot produces the AirJet Quanti [10, 12], which is a unique technology for the quantitative deposition of fluids or particulates, either for impregnation or for production of fine lines or spots on a substrate. A variant of this technology is produced by Kinematics Automation, Inc. (Twain Harte, CA) [13]. A schematic of the AirJet Quanti is shown in Fig. 8.6. This dispenser consists of a pneumatic spray generator hydraulically pumped with a positive displacement system such as a syringe pump. The typical generator is a pneumatic-driven aerosol similar to the artists' airbrush. In this case the dispenser/generator provides a quantitative delivery of a spray in terms of microliers per second, which when synchronized with motion can produce both individual spots and lines. Spots or lines can be made over a range of 0.5–25 mm in dimension. One of the major issues with the AirJet type of system is that the fluid delivery to a substrate has a Gaussian distribution and hence does not provide sharp edges to formats such as lines on a membrane. It is, however, a very important tool for coating/impregnation of surfaces and porous materials. This is exemplified by the use of impregnation for pad treatments and addition of the conjugate solution to the conjugate pad. In this case, the pad materials show sizeable variations of chemical concentration along the length of the pads due to variations in pad properties. This problem is

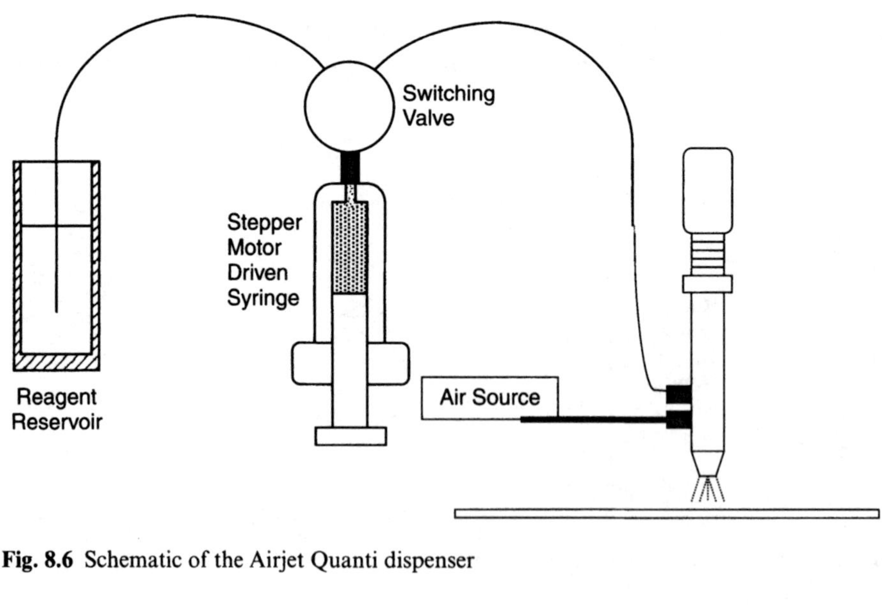

Fig. 8.6 Schematic of the Airjet Quanti dispenser

compounded in the batch drying process due to redistribution of reagents in the pad during drying. Batch drying ovens, by their very nature, also cause different drying conditions among positions within the oven. These factors result in an inherent variation in test properties using the impregnation processes. One solution to mitigate these problems for the conjugate process is the use of the positive displacement pumping of an airbrush dispenser (AirJet Quanti Dispenser). In this case, the AirJet Quanti can be used to imprint a quantitative strip of conjugate solution along the conjugate pad such that each test strip now has the same total concentration of conjugate per test with a CV of better than ±1%, dramatically improving the LFIA tests [3]. Further improvements could be made with better materials and pretreatment processes, which will be discussed later.

8.2.1.2 Impregnation

Impregnation is the saturation of the material with a reagent followed by drying. This can be done by soaking sheets or webs of materials in a reagent or using a dispenser to saturate the material. Either process can be performed using sheets or webs in various formats. A schematic of an in-line impregnation process is shown in Fig. 8.7. Typical applications of this method are the pretreatment of conjugate or sample pads with solutions of polymers, surfactants, and proteins to make them more hydrophilic, to control flow rates, or to impregnate them with sample treatment buffers. Blocking reagents are also typically applied to membranes using impregnation methods. The basic process involves the immersion of the material, either in pad or roll format, in a vat of the fluid to be applied. Many of the materials used in these assays require a somewhat lengthy dwell time to ensure complete impregnation of the material with the fluid, due to the extreme hydrophobicity of the materials (e.g., glass fiber or polyester pads). These processes may be either batch or continuous in nature. If they are batch, pads or sheets are immersed in the solution to be

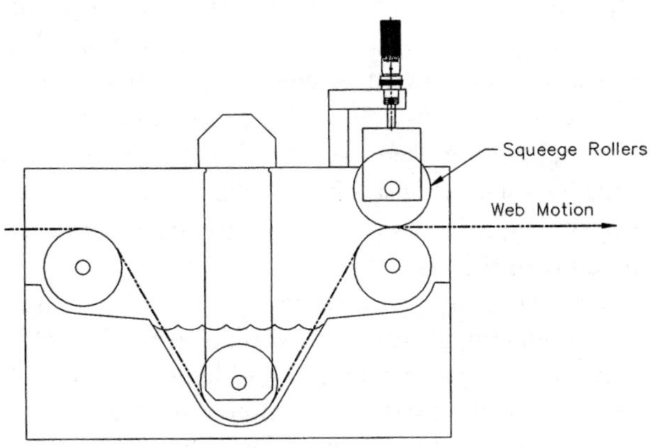

Fig. 8.7 Schematic of in-line impregnation

applied, agitated by hand or via another method such as ultrasonic vibration to ensure even uptake of the fluid. They are then removed, blotted, or squeezed to remove surface moisture, and dried using lyophilization or forced air at high temperatures. In a continuous process, the material, still in roll format, is passed through the bath at carefully controlled speed and contact angles, then squeezed and dried, usually in a drying tunnel or tower.

This process of immersion and drying, particularly in batch format, is a major source of variation for a number of reasons. First, uptake of fluids to the materials is not even. It is not possible to control the volume of fluid or the concentrations of dissolved components in the fluid absorbed into the material. Second, drying of the fluids is not even or consistent.

An alternative to impregnation by immersion is the use of a quantitative dispenser, such as a BioDot AirJet Quanti, to coat a strip or web followed by in-line drying. This method can reduce some of the sources of variation, by adding a defined volume of fluid to the material at a constant concentration of dissolved components.

8.2.1.3 Drying

Drying is one of the most critical processes to achieving low CVs and good sensitivity in LFIAs. There are multiple drying processes, with different aims in each case. For example, while drying test and control lines onto membrane after dispensing, the aim is to end up with a stable coating of immuno-active protein. As a result, the drying process is critical to the ultimate performance of the test, and must be very carefully controlled. There are several stages in the drying of proteins to the surface of a nitrocellulose membrane. First, the protein must partition from solution to the surface. Then binding to the surface must occur, initially by electrostatic interactions. Then, as water is removed and the hydrophobic interior of the protein unfolds and is exposed to the hydrophobic nitrocellulose, the hydrophobic binding process takes over. Finally, reorganization of the protein occurs. It is primarily this reorganization that defines whether the protein will remain active. The degree of drying, the drying method, and timing are all critical to achieve binding of a dried, active, and stable protein on the surface. The same degree of control is critical when drying particulate conjugates, such as colloidal gold or monodisperse latexes conjugated to proteins. Obtaining even drying, with the correct degree of moisture removal, is critical to achieving stability of the dried conjugate and facilitating even rewetting of conjugate pads and resuspension of the conjugate particles when contacted with the sample.

Two basic drying processes are in use by the LFIA industry, namely forced air at high temperature and lyophilization. Lyophilization is inherently a batch process with low throughput and hence, in spite of its superior drying properties, is not often used. The common approaches for forced air drying are batch ovens or in-line drying ovens for web formats.

8.2.1.4 Slitting

In the context used here, slitting is a cutting process used on materials along the web axis of the laminate (Fig. 8.1). Cutting, on the other hand, denotes cutting perpendicular to the web axis to form individual test strips. An important point to note is that both the slitting and the cutting processes can have important effects on the final test strip performance due to the influence of the material slit edges on the flow and resolublization of reagents during the test strip development. Slitting is principally used to cut wide sheets or web stock of sample and conjugate pad materials into thinner webs for final lamination as shown in Fig. 8.1. There are two basic strategies used in the industry: (a) slit untreated materials before reagent processing and (b) slit after reagent treatment. Since the drying process leads to redistribution of reagent to a material edge, elimination of dried cut edge will provide better reagent flow and resolublization.

8.2.1.5 Lamination

The lamination process refers to the process of bonding either strips or webs of the different materials to a pressure-sensitive adhesive on a plastic backing surface. The plastic backing comes with a protective release liner, which is removed prior to lamination. This release liner can be kiss cut to allow the sequential removal of selected parts of the release line for in-line lamination.

8.2.1.6 Cutting

The cutting process is used to cut the final test strip from the laminate either for bottling or for assembly into a plastic cassette. There are three types of cutters: (a) guillotine, (b) single rotary blade, and (c) rotary card cutter. The guillotine cutter cuts one part per cut cycle and can be used for high-speed cutting of laminates or for cutting individual test strips for pick-and-place assembly operations. The rotary card cutter gang cuts a laminate strip simultaneously into test strips at high speeds. This type of cutter can be used to cut and bottle strips or cut to bulk for either manual or other automatic operations. The rotary blade cutter is used for pick-and-place operations for automatic assembly operations.

8.2.1.7 Cassette Assembly

The typical test format is a test strip with a plastic housing, which can also include a desiccant. The test strip is placed into the bottom part of the cassette manually or automatically. Then, the top cassette cover is welded or snapped to the bottom part to complete the assembly.

8.2.1.8 Packaging

This operation consists of packaging the cassette in a foil pouch with a desiccant and, in some cases, other materials associated with the test. Again, packaging can be done manually using preformed pouches. The operator places the cassette and other parts in the pouch and then seals it with a rotary sealer. Packaging can also be done automatically where the pouch is formed in line with the other operations.

8.2.1.9 Environmental Control

For reagent processing, the environment needs to be at room temperature with reasonable temperature control and with humidity controlled in the range of 40–60%. A humid environment aids hydration of the materials and makes it easier for reagent incorporation into the different materials. After drying, the materials and final laminates need to be maintained at low humidity in the range of 20% RH (relative humidity) with constant temperature prior to packaging.

8.3 High-Throughput Manufacturing Processes for LFIA

In this section, dispensing, impregnation, drying, slitting, and lamination will be covered under the heading of "Laminate Assembly" and cutting, cassette assembly, and packing as "Final Device Assembly". Laminate Assembly includes the most critical processes that control device performance.

8.3.1 Laminate Assembly

A schematic of a laminate assembly is shown in Fig. 8.1. Currently, this configuration is made using two basic assembly approaches: batch processes that require a high degree of manual labor input and in-line processing which minimize this manual labor component. These approaches are shown schematically in Fig. 8.2. The dominant method used has historically been the batch mode for both R&D and manufacturing primarily due to the low capital cost and ease of transfer from a low-volume development environment to a manufacturing environment. In recent years, there is a trend toward using the in-line approaches due to the demand for increased production volume as a result of consolidation in the industry. It has also been recognized that the in-line approach can produce products that are much less variable than the more manual batch approach. Important features of the two manufacturing processes are outlined below:

i) *The Batch Process*

- It utilizes strip format for dispensing and lamination with strip length ranges from 150 to 500 mm.
- It processes strips in groups.
- The process time is non-critical.
- It provides poor process symmetry.
- It offers high degree of process flexibility.
- The process repeatability is operator-dependent.
- Automatic QC inspection method is difficult to incorporate.
- It is suitable for research and development use or for low volume of production up to 2–4 million tests per year.
- It requires high amount of labor.

ii) *The In-Line Process*

- It utilizes roll format for dispensing and lamination with 50–100 m roll.
- Rolls are used for continuous process.
- The process time is critical and limited by web speed and equipment size.
- It offers less flexibility in processing, but some flexibility can be accommodated by modular design of machine.
- There is excellent process symmetry.
- There is excellent repeatability of processes.
- It can be used from research and development phase to high-volume production phase with greater than 2 million tests per year.
- It is easy to incorporate automatic QC inspection on real-time basis and achieve continuous monitoring.
- It has low labor content.

8.3.1.1 Batch Processing

Batch processing is usually performed with a number of tabletop instruments that include: (1) an XY motion system with dispensers, (2) impregnation tanks, (3) drying ovens, (4) manual or semi-automated laminator, and (5) cutting methods for reducing roll or sheet stock to appropriate lengths and widths for lamination. Examples of these types of instruments are shown in Fig. 8.8. The XY motion system is typically equipped with two line dispensers for spraying test and control lines and one conjugate dispenser.

The batch method is basically the same for both R&D and manufacturing and can be scaled to higher-volume manufacturing by using duplicate systems. There are two major issues with the batch approach. The first is that quality is heavily influenced by manual operations. The second is that the batch process cannot be directly scaled to in-line processing. The major advantages are lower capital cost and ease of use. From a practical point of view, the batch process

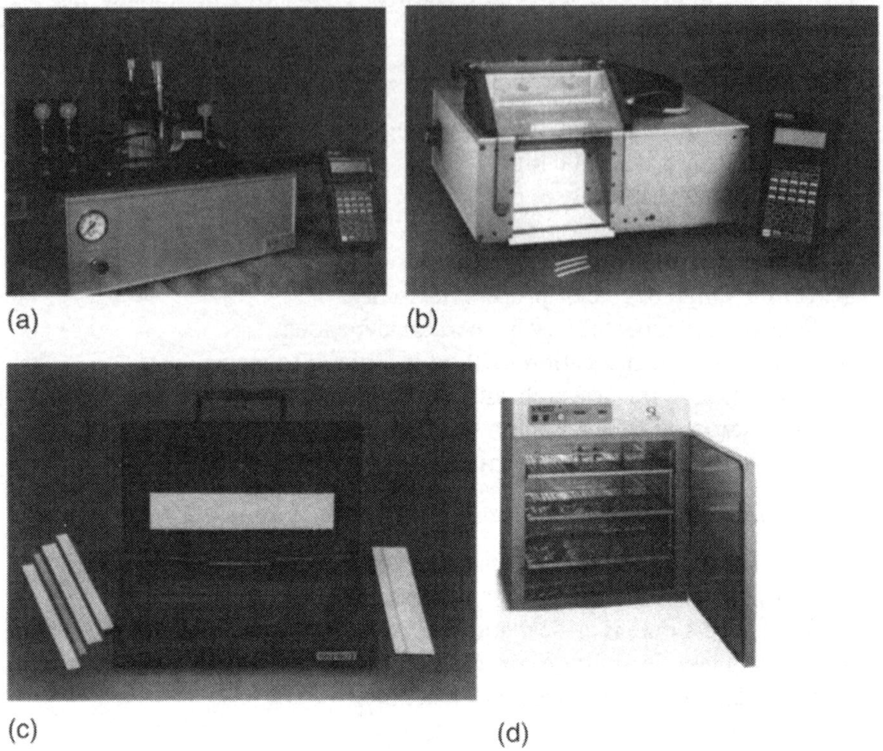

Fig. 8.8 Batch equipment

has served the LFIA well for test formats that are yes/no or threshold readouts manufactured in relatively low volumes (less than 10 million parts/year).

8.3.1.2 In-Line Processing

In-line manufacturing is a web-based processing where all materials are provided in a roll format. The in-line equipment is generally represented by three types of modules: (1) reagent processing and drying, (2) slitting reagent where webs are processed to narrow widths for lamination, and (3) lamination. Examples of these types of equipment are shown in Fig. 8.9 and described as below.

8.3.1.3 In-Line Reagent Processing

This type of machine is modular in design and can be configured to meet specific process requirements. The in-line reagent processing machine as shown in Fig. 8.9a consists of a number of modules. It is intended to multi-task between both dispensing and impregnation applications, and has capacity for in-line drying and automated part QC monitoring.

(a)

(b)

(c)

Fig. 8.9 In-line processing equipment

i) *Web Control and Dispense Module*: This machine consists of web feed and web take-up systems with tension and web speed control. It also includes web tracking systems, which are used to precisely locate the web edge relative to the dispenser positions and to align it to a common position on the take-up reel. The tracking systems are required to counter the camber that is inherent in roll stock of materials that are knife slit from master roll stock. This camber results in the drift of the web edge relative to a fixed reference point of 2–5 mm over the length of a 50 m roll. Taking the example of test and control line dispensing on nitrocellulose, such drift may cause the line position be offset from the membrane edge by as much as 2–5 mm. This displacement can be a source of increased product CVs and line position offsets in the plastic cassette.

Different dispenser types can be mounted to this control system, which is driven by tandem pumps. The tandem pump is a configuration of two syringe pumps that are connected to one dispenser and work in an offset mode to alternately fill and dispense to provide a constant dispense output over long web lengths. This module also accommodates a camera system (as shown) for inspection of dispensed lines. Lines can, for example, be assessed for continuity, position relative to an edge, and position relative to each other. Regions of the line dispense that do not meet inspection criteria are marked with a visible ink for rejection when the laminate is cut for final assembly.

ii) *Dip Tank Module*: This module provides a reagent reservoir with a roller system to impregnate a moving web. The module includes a refill system controlled by a reagent-level sensor. The refill system maintains a virtually constant slow flow of fluid into the tank. Such a system ensures a much more consistent solute concentration than would be the case in a batch tank, where solute drag is relatively uncontrolled. This module can also be provided with an enclosure with dedicated fume exhaust system.

iii) *Dry Tower Modules*: The dry towers are composed of a vertical web path to minimize the footprint or length of the composite machine. The dry path can also be designed in a horizontal configuration for dry formats that cannot be accommodated with the vertical design. The dry path as shown is 6 feet in length for each tower and is composed of six individually controlled heater zones. The air input for the convection drying process is supplied by three separate fans, each with variable speed control. The air flow for each fan has an independent flow path that directs the air to both sides of the web and then exits through a vent at the top of the dry tower. In this manner, dry air is continuously fed to the moving web surfaces throughout the web path length. The dry path and hence the dry time is increased by adding additional dry towers. The control temperature range is from room temperature to 80°C. Within the dry path, there are positions for non-contact infra-red (IR) temperature sensors, which can measure the web surface temperature. The principle of drying a wet material suggests that the temperature on the material would not reach the air temperature until the moisture has been removed due to the endothermic nature of the evaporation process. This provides the ability to use high dry temperatures during the initial drying process and then to reduce the temperature into the range of 40°C when the web temperature approaches the convection air temperature. In this way the kinetics of the drying process can be maximized.

8.3.1.4 Lamination or Cutting Modules

The module design of the in-line reagent system provides the ability to add other types of process modules for value-added operations. With these modules, it is possible to cut the web into strips rather than re-roll or to laminate the reagent-treated webs to other materials such as adhesives with release liners or plastic backing. Another possibility is to laminate membrane to plastic backing, followed by dispensing of the test and control lines.

8.3.1.5 In-Line Slitting

This module is shown in Fig. 8.9b and is used to slit wide webs into smaller width web formats. The machine consists of a reel feed system, a rotary blade system, and a number of take-up reels. The take-up reels are designed to be transferred directly to the in-line lamination system discussed later. The rotary blade system is composed of a series of cutting blades with spacers so that the cut width can be customized by using different width spacers. An example would be to use a 100 mm wide web of the conjugate material and use nine airjets to spray nine conjugate lines. The web is then dried, re-rolled, and taken to the slitter and slit to nine smaller webs 10 mm wide for use on the in-line lamination machine. Impregnated webs would also be processed as 100 mm wide webs and then slit to the appropriate lengths for automatic lamination.

8.3.1.6 In-Line Lamination

The function of this system is to laminate the assembly (Fig. 8.1), using web formats of all the various layers that have been pretreated and slit to the appropriate widths. These webs are laminated to a plastic backing with precut release liners. Again, such a system is designed as a module system to allow for a wide range of laminate structures. An example of an in-line lamination machine is shown in Fig. 8.9c. This machine consists of the following modules:

i) *Backing Feed*: This module consists of a feed reel with tension and speed control and a take-up reel for the membrane-release liner.
ii) Membrane Lamination: This module consists of a feed reel for the membrane with tension control, a take-up reel for the membrane liner (if required), and a lamination roller. Also included is a web tracking system, which senses the membrane edge and aligns it to a position on the plastic backing. This station can laminate both backed and unbacked membranes.
iii) *Multiple Material Feeds*: This module includes three roll feeds for the conjugate, sample, and absorbent pads with lamination rollers. Also included are take reels for the various release liners. Mechanical guide systems are used for referencing the webs to the plastic backing.
vi) *Vision Inspection/Bad Mark*: A vision system is used to inspect the different laminate placements and positions. Out of specification are ink marked for removal at the cutting process.

v) *Re-Roll/Cut*: This module either cuts the laminate into strips or re-rolls for feeding the strip cutting operation. The re-roll requires a large-diameter core to prevent stress damage to the laminate.
vi) *Other Options*: Other modules can include adding plastic overlays to the top of the laminate or performing dispensing operations at some point in the lamination process.

8.3.1.7 Throughput Considerations for Laminate Assembly In-line Systems

For throughput calculations, one can consider the process layout shown in Fig. 8.2, which utilizes five reagent processes and a single lamination process to laminate the membrane, sample pad, conjugate pad, and absorbent pads to the plastic backing. It can further be assumed that the conjugate and sample pads are processed as 100-mm wide webs, which, after reagent treatment and drying, are slit to the appropriate width for lamination. Using the in-line manufacturing machines as described above, one can estimate the manufacturing throughput as follows:

i) *Reagent Processing*: With the appropriate drying path lengths, each of the reagent processes can be performed at a rate of about 15 parts per second for a final test width of 5 mm. Such a calculation is based on the assumption that the 100-mm wide conjugate and sample pad webs can be processed with reagent applications and dried at a rate of about 10 mm per second and each 100 mm web will yield about eight smaller webs for lamination after slitting. For a test that is 5 mm wide, this would produce about $2 \times 8 \times 0.8 = 12$–13 parts per second where 0.8 is a utilization factor for set up. If the membrane processes can be done at rates of about 8–12 parts per second (50–75 mm per second web speeds with 0.8 utilization factor) using a single in-line reagent system for five processes, this would yield an average manufacturing throughput of about 2–3 parts per second. For a single 8-hour shift, about 10 million parts per year can then be made. If dedicated in-line processing machines are used for each of the reagent applications, the throughput would increase by a factor of 5. By working additional shifts, the throughput for a dedicated product can be increased to greater than 200 million parts per year.
ii) *Slitting*: The slitting machine as shown in Fig. 8.9b can slit the reagent-treated webs at a rate of about 100 mm per second, which equates to about 160 parts per second. Thus, one slitter can easily accommodate production rates, for the product described in (i), in excess of 200 million parts per year.
iii) *Lamination*: The laminator shown in Fig. 8.9c can operate in the range of 100 mm per second, which equates to about 20 parts per second in-line speed or about 15 parts per second considering setup times. The laminate can be cut into strips or taken up on large-diameter reels. This type of machine can operate in the range of 40–50 million units on a per shift basis.

8.3.2 Final Device Assembly

The final assembly is the most operation-intensive part of the LFIA product process. It is labor-intensive if manual assembly is utilized and capital-intensive for automatic assembly. This problem is compounded by the fact that the plastic housing design is usually customized for each individual product, resulting in the need for special tooling.

i) *The Batch Process*: In the batch mode, manual assembly requires various assisting tools for cutting, assembly, and packaging. The equipment needed here are cutters, tooling for closing the plastic housing, and a sealing machine for use with preformed foil pouches. The parts are cut using high-speed guillotine or rotary card cutters where the cut parts are collected in bulk. Operators then individually place the cut strips into the bottom of the plastic housing, put on the plastic top, and use a pressure tool to snap the two parts together. An operator then places the cassette in a foil pouch with a desiccant prior to sealing it with a manual rotary sealing mechanism. This approach requires the least capital but is most labor intensive. It is also the worst scenario from the quality control point of view, and has the potential to worsen with increasing volume and number of operators. In addition to being the most susceptible to operator error, this step also involves the maximum number of part-handling operations, which can both damage parts and impart a level of variability due to different operator judgment. In the batch process, part damages and operator errors can further be amplified during the cutting operations where cut parts are handled in bulk and then manually picked for assembly. Thus, for LFIA formats demanding high quality and/or quantification, a fully batch approach should not be used.

For high quality and/or quantitative test formats, a minimalist approach would be to use an assembly work station where laminate cards or roll stock are fed to a cutter that places the cut strip in the plastic part as shown schematically in Fig. 8.10. The cutter system should have ink mark sensors that recognize parts on the laminate which have been error marked upstream. These parts are rejected at the cut operation. Once the test strips are placed within and protected by the plastic housing, the device is more amenable to manual handling. The cycle time of this type of machine is in the range of 6–8 seconds or less than 1 millions parts per year per shift.

ii) *The In-Line Process*: A full in-line final device assembly operation would include a machine similar to that shown in Fig. 8.10, but using feeder bowls instead of operators. These types of machines can be made to operate in the 1–4 seconds per part level and includes rejection of bad strips at the cutting station with throughputs in the range of 2–6 million parts per year per shift. The assembled cassette would be output to an in-line pouching machine and then onto a boxing machine. The assembly and pouching machines should be located in a relative dry environment in the range of 20% humidity. The latter two machines are readily available from companies that supply such equipment to

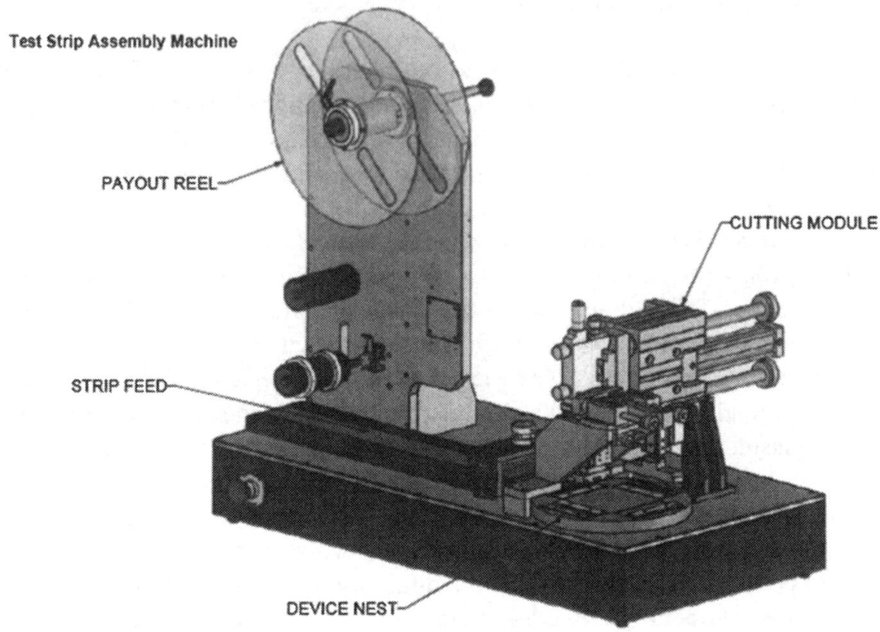

Fig. 8.10 Schematic of an assembly work station

the food and related industries. A single system that does assembly, packing, and boxing at the rate of 6 million parts per shift per year will be three to five times the cost of a laminate assembly machines described above to achieve the same throughput.

8.3.3 Conclusion

The manufacture of LFIAs is an inherently complex process with numerous sources of variation built into the product. Many of the manufacturing steps are employed to help minimize some of the variability. Manufacturing processes must be carefully designed and tightly controlled to minimize the introduction of further variability into product performance. The typical processes and the main sources of risk to finished product variability are identified in Table 8.2.

From the discussions above, some conclusions on how to generate better performance characteristics using existing technology can be defined:

1. Move away from batch processing to in-line processing
2. Utilize automated in-line process controls for dispensing, drying, and lamination
3. Dispense membrane reagents using non-contact methods
4. Replace immersion reagent methods with quantitative dispensing methods

5. Quantitatively dispense conjugates
6. Eliminate reagent induced treated edges on the sample and conjugate pads with a slitting process
7. Eliminate the use of cutting processes that use bulk handling of test strips
8. Automate lamination and assembly processes.

Table 8.2 Summary of LFIA processes and methods currently used

Component	Process	Method	Risk
Analytical membrane	Dispensing of test and control lines		
	Blocking	Immersion	Non-uniform concentration
	Drying	Forced air at high temperature	
Conjugate pad	Pretreatment	Immersion and blotting	Non-uniform concentration
	Application of conjugate	Dispensing	
		Immersion	Non-uniform concentration and release
	Drying	Forced air at high temperature	
		Lyophilization	Requires a non-uniform batch process
	Slitting	Rotary slitter	Web edge influences flow in test strip
Sample pad	Pretreatment	Immersion and blotting	Non-uniform concentration
	Drying	Forced air at high temperature	
	Slitting	Rotary slitter	Web edge influences flow in test strip
Laminate (card or roll)	Lamination of components	Manual	Handling can damage parts
		Auto Laminator	
Strip	Cutting into individual strips	Guillotine	Damage to test strip resulting in non-uniform flow
		Rotary cutter	Same
	Placement in cassette and/or packaging	Manual	Damage to test strip
		Pick and place	
Overall device	Environmental control	Humidity control	Can effect shelf-life time

8.4 Next-Generation Requirements for Manufacturing Tools for LFIA

8.4.1 Materials and New Product Designs

In recent years, a tremendous amount of innovation has developed in the PON immunoassay segment, driven by patent pressures and a need to generate assays with improved sensitivity, reproducibility, quantization, and objective read/record technology. Some of the development effort in innovative technologies is designed to displace existing technologies, such as lateral flow, in the medium to long term. There are, however, elements of the LFIA that make them extremely attractive both to existing markets and to those currently evolving. LFIAs are still relatively low cost, easy to use, and, importantly, enjoy acceptance and recognition from users and regulatory authorities. It has been repeatedly shown that the introduction of completely novel displacement technology into the market, particularly the clinical diagnostics area, requires decades of work and hundreds of millions of dollars. As of today, there are not many displacement technologies close to being ready to take on this task. As a result, there is a strong urge to maintain the best elements of lateral flow technologies and overcome the basic performance issues that have hampered its broad application in highly sensitive and highly reproducible applications.

In order for lateral flow-based assays to evolve into technologies that can address the criteria demanded by the next-generation diagnostic markets, several fundamentals need to be addressed. Primarily, the assays need to be made more reproducible, more sensitive, easier to manufacture, easier to operate and interpret, and, from a clinical point of view, provide results that are of relevance and that correlate with other laboratory-based diagnostic systems. Several elements are being addressed in the marketplace today to achieve these aims. These include the following:

- Improved control of manufacturing processes
- Application of material science to bring novel, custom-designed, and more appropriate materials to bear on POC assay systems
- Application of new labeling and reading technologies
- A much more widespread acceptance of the need for outsourcing of critical services as projects become more multidisciplinary in nature [3].

From the perspective of basic manufacturing principles, several assay and material design approaches are showing promise. Improvements in materials that avoid the need for some of the more variable manufacturing processes, primarily impregnation and drying, include the generation of custom materials capable of multiple functionalities currently embodied in different materials. This approach removes the need for individually treating and laminating separate pads, which in turn removes several sources of variability. One of the weaknesses in the design of the LFIA (as shown in Fig. 8.1) is the mechanical interface between the sample and conjugate pads and that between the conjugate pad and membrane. These mechanical interfaces can be sources for flow variations and hence are sources for increasing CVs. The Fusion 5 material (see Chapter 7) produced by Whatman (London, UK) is an attempt to eliminate these two interfaces by using a single membrane to serve as sample pad, conjugate pad, and membrane [14]. Other design formats have used a single pad for both the sample and the conjugate pads. However, these approaches may present new problems for the pretreatments associated with sample and conjugates as a single pretreatment reagent may not give the best results. The combined use of one material for multiple flow processes also rules out the use of impregnation as a method of treatment. This problem can be resolved by using dispensing methods to apply pretreatment reagents in isolated areas on the same pad.

Another approach to remove the mechanical interference and material treatment issues is the incorporation of fluidic elements into assays. One example is provided by Amic AB (Uppsala, Sweden) [6], which has introduced a molded plastic device consisting of a micro-fluidic flow path with arrays of small pillars to generate the capillary force for flow. The Amic device is a discrete plastic chip the size of a glass slide. One advantage of this particular design is that the fluidic path is very reproducible. Theoretically, these characteristics should not change with time, which is one of the major sources of variation in a membrane-based assay. This assay format aims at applications requiring high sensitivity and quantization. Hence, fluorescence conjugates are typically employed. A number of assay formats have also been demonstrated. The shape of the micro-capillary fluid path allows the various reagents, such as the conjugate as well as the test and control lines, to be placed in discrete locations rather than in the format of continuous lines on a web. Also the tests are discrete parts that need to be handled individually. Thus, the manufacturing processes are quite different from those used to produce a membrane-based LFIA.

8.4.2 New Manufacturing Processes

The discussion here is limited to in-line manufacturing of the "Laminate Assembly" as this is where the major opportunity for improvement exists. We have discussed above that the use of quantitative non-contact dispensing, when combined with in-line processing, has made major strides to improving device CVs. In these cases, the dispensing was limited to test/control lines and conjugates. The next step is to implement more control elements for those process parameters that can influence device performance. Listed below are the items to consider for each of the machines in the "Laminate Assembly" domain.

8.4.2.1 Recommended Reagent Processing

- Replace all impregnation processes with quantitative in-line dispensing.
- Implement a controlled humidity environment for all dispensing operations in the range of 40–60% in order to enhance hydration of the materials, to reduce static charge, and to make the materials more hydrophilic.
- Implement a controlled humidity environment (<20% R.H.) for all roll stock after reagent and drying processing in order to protect the dried down reagents and inhibit denaturing of proteins.
- Improve monitoring and control of the drying process, which includes: (a) monitoring of web temperatures in the dry path; (b) monitoring of air flow, temperature, and humidity for the effluent from the dry tower vent; and (c) data logging of all these parameters to provide long-term analysis of data trends and device performance. The goal here is to control the extensive process properties, mass flow, and humidity in order to provide for highly reproducible process conditions.
- Consider a process for hydrating all web materials prior to processing. This can be done off-line. The goal here is to process materials that have a controlled level of hydration and consistent hydrophobicity throughout the reagent addition processes. This would include a re-reeling machine that adds leaders to each end of the roll stock and exposes the web materials to a constant humidity environment. The hydrated roll stock will then be stored in a constant humidity environment.
- Develop methods for managing reagents and roll stock for fast setup times and maintaining process conditions during roll stock change over. The management method will allow quick change-over of new reel stock with leader materials on each end of the webs to allow for quick start-up with minimal wastage of reel stock. Complete reel stock would be stored in a constant humidity environment in the 20% range, which could be achieved using a dry nitrogen chamber with a small positive pressure.

Figure 8.11 shows a schematic of an in-line reagent processing system with the above features.

8.4.2.2 Slitting

Since the materials to be slit have been pretreated and kept at a low humidity, the slitting chamber should also be kept under a low positive pressure of dry nitrogen. Slit rolls should be stored under the same conditions.

8.4.2.3 Lamination

Lamination processes need to be performed in a dry environment in the range of 20% RH. This can be done in a controlled room or controlled chamber surrounding the lamination process.

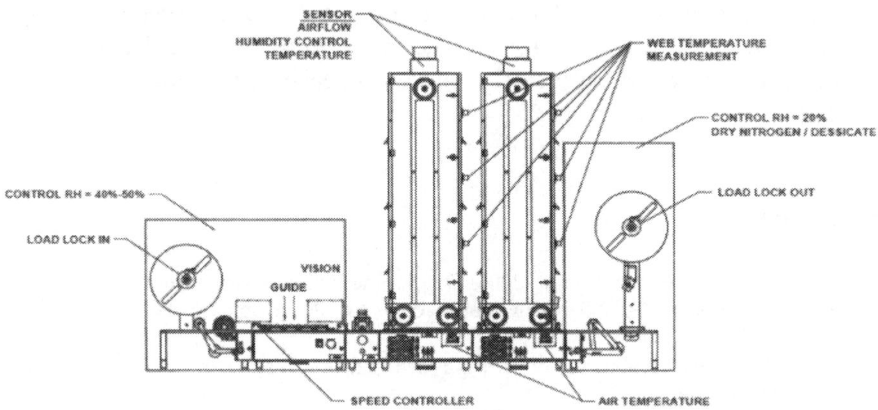

Fig. 8.11 Next-generation in-line reagent processing

8.5 Conclusions

As the in vitro diagnostic (IVD) industry undergoes a shift toward new technologies and new analytes, the relationships between the assay developer, the manufacturing process designer, the equipment supplier, and the manufacturer of the final product are becoming more critical to success than ever. The IVD industry is continuing to move toward novel solutions to performance and user-related issues, including miniaturization and multiplexing of assays, as well as the use of novel microfluidics, labeling, and reading technologies. Many of the newer analytical targets, particularly in the POC diagnostics markets, require easy-to-use and quantitative systems, which in turn demand more of the product in terms of reproducibility, stability, sensitivity, and dynamic range. As demands on the product increase, so do demands on the manufacturing process and the manufacturing process machinery. A natural corollary to the development of novel technologies and applications is the development of custom manufacturing processes and machinery to produce the final product. Customization brings unique challenges and increased risk to any project. As a result, it is critical that manufacturing process design be performed as early as possible in a design and development cycle.

The trend toward outsourcing of many elements of product development and manufacturing is another natural result of the increase in complexity of assay development processes. Development and production of newer-generation POC devices require input from a wide variety of specialist disciplines. Particularly for smaller companies, the range of skills required to complete the design of a complex IVD device can be difficult to maintain or access. This can also be true of larger corporations whose focus may be elsewhere. As device design in this area begins to incorporate more complex elements, such as integrated readers, the demands on the manufacturing technology become extreme. Great care must be taken in defining the specifications both for the device and the manufacturing process as early in the design process as possible.

References

1. O'Farrell, B. and Bauer, J. (Jun. 2006) Developing highly sensitive, more-reproducible lateral-flow assays. Part 1: New approaches to old problems. IVD Technology 12(5):41–49.
2. Tisone, T.C. (2000) In-line manufacturing for rapid-flow diagnostic devices. IVD Technology 6(3):43–60.
3. O'Farrell, B. and Bauer, J. (Jul. 2006) Developing highly sensitive, more reproducible lateral flow assays. Part 2: New challenges with new approaches. IVD Technology 12(6):67–75.
4. www.quantumdesign.com
5. O'Farrell, B (2008) Emerging technologies that facilitate the field use of rapid diagnostic tests. Proceedings of the SPIE Defense, Security and Sensing Conference, March 2008.
6. www.amic.com
7. www.chembio.com
8. Rosenstein, R. (Jan. 7, 1997) Solid phase chromatographic immunoassay. US Patent 5591645.
9. Charlton, D. (Nov. 26, 2002) Test device and method for colored particle immunoassay. US Patent 6485982.
10. Tisone, T.C. (Apr. 14, 1998) Precision metered aerosol dispensing apparatus. US Patent 5738728.
11. Tisone, T.C. (Apr. 28, 1998) Precision metered solenoid valve dispenser. US Patent 5743960.
12. Tisone, T.C. (Jun. 29, 1999) Dispensing apparatus having improved dynamic range. US Patent 5916524.
13. www.Kinematics.com
14. www.whatman.com (Fusion 5)

Chapter 9
Handheld and Portable Reader Devices for Lateral Flow Immunoassays

Konrad Faulstich, Roman Gruler, Michael Eberhard, Dirk Lentzsch, and Klaus Haberstroh

9.1 Introduction

In recent years, lateral flow immunoassays have become an invaluable tool for various diagnostics applications. Among the most prominent reasons for this development are their reasonable sensitivity and specificity for many applications and their rapid time-to-result readouts. The samples are applied to the test strips directly, often without the need of prior time-consuming sample preparation steps. Lateral flow immunoassays are also easy to operate and, last but not least, they do not require a device for readouts. Therefore, they are cheap and mobile. Like other technologies, however, lateral flow immunoassays also have limitations and do lack important features to further exploit this technology [1]. These shortcomings include lack of automated documentation, subjective interpretation of results leading to a high number of false positives and false negatives, lack of accurate quantitative and limited multiplexing capabilities, as well as limitation in high-throughput diagnostics due to manual operation. Sensitivity is limited by the K_d of the antibody–antigen conjugate and by the colorimetric readout using the popular colloidal gold beads technology. In addition, specificity is influenced by the cross-reactivity of antibodies (see Chapters 4 and 10), and this limits their applicability to "good" and stable compounds. To overcome some of the limitations, both readers as well as novel biochemical techniques are being developed [2–6] by many companies eager to differentiate themselves from rivals in this highly competitive industry and to offer higher-quality products combined with more convenience to the customer. In fact, many lateral flow test manufacturers who have or are planning to launch products based on a reader find it almost a necessity to offer a complete system including test strips and reader.

In this chapter, we describe the development of handheld and portable lateral flow readers and discuss the usefulness and specifications of these devices for rapid immunoassays. We point out how these devices help overcome

K. Faulstich (✉)
ESE GmbH, Stockach, Germany
e-mail: Konrad.Faulstich@ese-gmbh.de

R.C. Wong, H.Y. Tse (eds.), *Lateral Flow Immunoassay*,
DOI 10.1007/978-1-59745-240-3_9 © Humana Press, New York, NY 2009

limitations of current lateral flow immunoassay technology to enable further exploitation of this technique. We define portable readers as devices that are of laptop size or less, can be carried around by one person at all times, and have a weight in the range of 1–2 kg. For handheld readers, these are functional units that can be held and operated in one hand and have the appropriate size and weight less than the portable readers defined above.

9.2 General Considerations

9.2.1 The Voice of the Customers

Customers and users of lateral flow immunoassays report repeatedly that certain features are very important to them and need to be included in a lateral flow reader besides accurate performance. Examples of the most common customer requests and how these needs are translated into practical use in one commercially available reader are listed in Table 9.1.

9.2.2 Reader Devices

Lateral flow immunoassay test strips are established for more than 30 years. It is surprising that commercially available readers only appear in number in recent year. This is probably due to the fact that, during the early years of lateral flow immunoassay development, the focus was on low cost test to provide yes/no answers, whereas the focus today is on exploiting the technique for more sensitive and quantitative performance. This approach extends lateral flow immunoassay to new markets and applications not existed before. Although lateral flow immunoassays are very affordable and the readout can be performed visually without reader devices in colloidal gold- or latex bead-based assays (see Chapter 5), inconvenience in manual documentation and subjective interpretation often lead to false negative and false positive results. Such shortcomings of lateral flow immunoassay tests have created certain demand for automated laboratory systems. However, the available systems do not fulfill the features requested by the customer's needs (Table 9.1), especially on mobility, simplicity of operation, speed, low cost, and the avoidance of time-consuming sample preparation steps.

To improve lateral flow immunoassay sensitivity, some companies have replaced colloidal gold beads with fluorescence dyes and paramagnetic particles (see Chapter 5). Since fluorescence dyes cannot be detected by the naked eye directly, such technologies immediately create demands for the development of readers for quantitative analysis. The same applies to paramagnetic particle–based lateral flow immunoassays.

Just very recently, several reader devices have become commercially available and these can be grouped into fluorescence, magnetic particle, and

Table 9.1 Most common customer-requested features available on the *ESE-Quant* lateral flow reader

Customer request	Feature of *ESE-Quant* lateral flow reader
Ease and convenience of operation	• One-button operation • Delay time scanning (the reader starts scanning automatically only when the test is ready for readout – at a time x minutes after sample is applied to test strip) • Simple method menu allows scanning of different tests (up to 16) with the correct parameters • All results are displayed in an easy-to-read table format • Reader performs automatic self-test and self-calibration (not necessary to use a reference cassette or internal standards) • Field-based tests possible as readers can be used as stand-alone devices and run on rechargeable batteries • Car battery adaptor available
Quantitative read out	• High accuracy • Professional software algorithms • High dynamic range of measurement • Several calibration curves established as standard • User can enter any calibration curve • Data output in numbers with units, action based or qualitative
Automatic electronic documentation of results	• Records and saves data • Data uploadable to PC or USB Stick
Higher sensitivity	• Highly optimized sensor • Fluorescence reading
Objective interpretation of results	• Cut-off levels can be set by operator • Calibration curves can be entered to translate the data into meaningful results • Batch numbers can be input • Self-test and self-calibration of reader
Use of reader in QC for test strip manufacturing	• Quick scans allow high throughput • Batch numbers can be calibrated • Barcode reader and RFID systems available • Stand-alone reader • Portable • Different tests can be quality controlled due to simple test method menu
Handheld reader format or mobility	• Handheld and small portable readers are available
Operational robustness	• Place test strip into reader and scan • Write-protected data • Automatic self-test and self-calibration • Data get assigned a run number, date, and time automatically

Table 9.1 (continued)

Customer request	Feature of *ESE-Quant* lateral flow reader
Physical robustness	• Readers work within a high temperature and humidity range
Audio/visual display of results	• Good visible display • Beeper for confirmation or errors
Connectivity to PC, printer, barcode reader, or other data management system	• Internal or external barcode reader available • Internal or external RFID data management system available • USB port • Printer available
Hard copy of test results	• Connectivity to printer, small handheld thermal printer as accessory
Compatibility to clinical workflow and systems	• Connectivity to barcode reader • Internal or external RFID data management system optional
Stand-alone reader without use of computer	• Reader display • Big reader memory • Reader beeper • Real-time clock • Battery display • Runs from rechargeable batteries • Reader keyboard to enter information
Batch and calibration data management system	• Barcode reader • RFID tags • Download from USB stick or webpage • Manual entering of data • Preinstalled pull-down menu
Low price/investment	• Integrated flexibility: • Easily adaptable to any cassette format • Easily adaptable to any label
Appealing industrial design	• Convenient look and feel • Professional industrial design • Printable customized logo and reader model • Customized colors
Self-test/self-calibration of reader	• Included • No need to place a calibration or reference cassette
Fast read out	• Scans within seconds
Wireless data transfer/remote control	• Optional feature
Compatibility to customer's unique cassette format	• Standard feature
Compatibility to customer's unique label	• Standard feature
Availability of professional Software	• Standard feature (for assay development, reader configuration, and end-user operation)
Compatibility to different tests and test formats	• Standard feature • Software and firmware allow test specific reader configuration
Multiplexing	• Standard feature • Up to 16 test lines (more upon request)
Full customization	• Fully customized hardware and software

colorimetric readers. In many cases, charge-coupled device (CCD)-based image systems and scanning systems are also utilized in these readers.

9.3 Reader Device Systems

9.3.1 CCD-Based Imaging Systems

CCD-based imaging systems have the advantage that they can provide an image of the entire lateral flow immunoassay strip, similar to what the customer is used to see from visual read-outs of colorimetric stripes. This is an important factor for customers whose lateral flow strips do not perform optimally, resulting in the non-homogeneity of the bands on the lateral flow strips. For these reasons, some test manufacturers and marketers may recommend to customers that the "bad parts" of the test and control bands be excluded from the analysis, which is possible using CCD-based systems and imaging software. However, this is a questionable practice because the "good parts" of the test band is not representative of the whole test. Such tests should be evaluated as invalid. Low-performing tests are not addressed in this paper and should not be considered for use with reader devices. Besides the chemistry, other effects that may influence the quality of the band such as illumination and lens effects also need to be considered. Even the best homogeneous band may appear distorted if the illumination is not uniform. This is particularly true for fluorescence-based read-outs. Another advantage of imaging systems is that there are no moving parts. Moving parts may break and lead to inaccuracies if the devices are not engineered properly. On the other hand, CCDs are large, heavy, and expensive. These features are often incompatible with a truly handheld/portable lateral flow reader.

In addition, the number of data points that need to be recorded in CCD-based systems is very high, especially when a reasonable resolution is required. This may result in very limited internal memory capacity in the systems or make the system quite expensive if the memory is upgraded. In this scenario, a computer is always required. Therefore, most CCD-based systems are neither handheld nor mobile unless a "Palm" or pocket PC is attached to it. However, due to the rapid advances in the computer industry, pocket PCs and palm-size computers change very quickly and parts may become obsolete. A system based on a commercial pocket PC can be outdated after only a few months and the vendors will not be able to deliver the same instrument to the customers, unless they buy a high volume of the components or devices and take the risk of not being able to sell them all, resulting in upward pressure on price. The alternative to this scenario is that the customers have to accept an ever-changing instrument. In a regulated medical diagnostics market, this is the worst-case scenario since the customers need a locked-in design in order to obtain regulatory clearance.

There are also limitations related to the miniaturization of CCD-based systems, as the camera needs a certain field of view to capture the image from the entire strip. Perhaps, the most important point to consider is the price sensitivity of the lateral flow immunoassay market. CCD-based systems contain naturally expensive components and may be price prohibitive to customers. Some vendors may provide the devices to their customers for a low price. Others may charge only the cost of manufacturing or even give them away for free in some cases, hoping to recover the costs in more rapid turnover of high-margin consumables. A solution to the price and miniaturization issues may come from lower cost complementary metal oxide semiconductor (CMOS)-based systems. These systems, nevertheless, do not perform as well.

9.3.2 Scanning Systems

Scanning systems have the advantage that the scans are performed very rapidly, including data evaluation. No imaging software and processes need to be performed. Memory size can be much smaller and more scans can be stored on the device directly. This opens the possibility to operate and perform tests completely independent of any connections to computers and provides a simple solution to field-based tests and truly handheld devices. This simplicity is also reflected in the price, which is lower. Moreover, customers do have the option to connect those devices to a computer for further data storage and analysis.

The disadvantage of the scanning systems is that no image of the test strip is available. In addition, the resulting data look different from the images the customer is used to see from visual read-outs. However, most lateral flow immunoassay customers do not request an image as data output. They request either a qualitative data output in the formats of positive/negative/invalid or a quantitative output such as numbers and units or even just a recommended action. These interpreted data outputs are meaningful to the operator of the instrument and test users. The recording of data is seen only as an intermediate step by the customers.

9.3.3 Other Systems

There are also devices available, which are neither CCD-based nor scanners. One example is the so called Fluorescence Visualizer [7]. Fluorescence cannot be read by the naked eye directly. Therefore, an illumination source and a filter are required to read the results. Such devices are very affordable and allow the customer to read out the highly sensitive fluorescence lateral flow assays visually as is the case with colorimetric lateral flow assays.

Table 9.2 lists some commercially available lateral flow readers and selected features. There are various approaches to the manufacturing of readers and these vary from colorimetric, fluorescence, to magnetic devices. They can also be ultra-small handheld or large-sized desktop readers. Although this paper

Table 9.2 Examples of commercially available lateral flow immunoassay readers and selected features

Reader model	Marketer	Reads different cassette/ strip formats	Reads different labels	Reads Tests from various manufacturer	Both stand-alone reader and PC-controllable	Label	Accessories/ connectivity	Size	References
ESE-Quant	ESE GmbH (OEM* partner)	Yes	Yes	Yes	Yes	• Gold beads • Colorimetric/ Reflectrometric (any dye) • Fluorescence (any Dye)	• Barcode Reader or RFID Data Management System • Printer • Charger • USB port to PC • Car battery adaptor • Suitcase • Wireless (optional) • Software/Firmware • Full customization	Small Portable Desktop/ Handheld	[7]
GoldScan	ESE GmbH (OEM* partner)	Yes	Yes	Yes	Yes	• Gold beads • Colorimetric/ reflectrometric (any dye)	• Barcode Reader or RFID Data Management System • Printer • Charger • USB port to PC • Car Battery • Adaptor • Suitcase • Wireless (optional) • Software/firmware • Full customization	Small Handheld	[7]

Table 9.2 (continued)

Reader model	Marketer	Reads different cassette/ strip formats	Reads different labels	Reads Tests from various manufacturer	Both stand-alone reader and PC-controllable	Label	Accessories/ connectivity	Size	References
FluoScan	ESE GmbH (OEM* partner)	Yes	Yes	Yes	Yes	• Fluorescence (any Dye)	• Barcode Reader or RFID Data Management System • Printer • Charger • USB port to PC • Car Battery Adaptor • Suitcase • Wireless (optional) • Software/firmware • Full customization	Small handheld	[7]
Fluo Visualizer	ESE GmbH (OEM* partner)	Yes	Yes	Yes	No	• Fluorescence (any Dye)	N/A°	Ultra-small Handheld	[7]
Triage Meter Plus	Biosite/ Inverness	No	No	No	No	• Fluorescence	N/A°	Desktop	[9]
Cholestech LDX	Cholestech/ Inverness	No	No	No	No	• Colorimetric	N/A°	Small Desktop	[10]
Clearblue	Inverness	No	No	No	No	• Colorimetric	N/A°	Ultra-Small Handheld	[11]
RapiScan	Cozart	No	No	No	No	• Colorimetric	N/A°	Handheld	[12]
Diagnostic Test Reader	Alverix	Yes	Yes	Yes	N/A*	• Colorimetric • Fluorescence	• customization	Handheld/ Desktop	[13]
RAMP	Response Biomedical	No	No	No	N/A*	• Fluorescence	N/A°	Desktop	[14]

Table 9.2 (continued)

Reader model	Marketer	Reads different cassette/ strip formats	Reads different labels	Reads Tests from various manufacturer	Both stand-alone reader and PC-controllable	Label	Accessories/ connectivity	Size	References
ROSA	Charm Sciences	No	No	No	N/A*	• Colorimetric	• Barcode reader • Printer • Car battery adaptor	Desktop	[15]
Defender TSR	Alexeter Technologies	No	No	No	Yes	• Colorimetric	• Wireless • USB to PC • Software	Handheld	[16]
AccuScan	Neogen	No	No	No	Yes	• Colorimetric	• Connectible to PC • Software	Handheld	[17]
Mobile Reader 500	Matest	Yes	No	Yes	Yes	• Gold Beads	• Wireless • Software	Desktop	[18]
LabU-reader	77 Elektronika Kft.	No	No	No	Yes	• Gold Beads	• RS 232 to PC • Barcode reader • High throughput • Software	Large Desktop	[19]
HandU-Reader	77 Elektronika Kft.	No	No	No	Yes	• Gold Beads	• RS232 to PC • Docking station • Software	Desktop/ Handheld	[19]
MICT MAR	Magna Biosciences	No	No	Yes	N/A*	• Magnetic Beads	N/A°	Handheld	[20]
MIATek tabletop	Magnisense	No	No	No	Yes	• Magnetic Beads	• PC connection	Desktop	[21]
Cardiac Reader	Roche	No	No	No	N/A*	• Gold Beads	N/A°	Portable	[22]
Immuno-chromato-Reader C10066	Hamamatsu	Yes	Yes	Yes	Yes	• Colorimetric	• RS 232 to PC • Software	Desktop	[23]

Table 9.2 (continued)

Reader model	Marketer	Reads different cassette/ strip formats	Reads different labels	Reads Tests from various manufacturer	Both stand-alone reader and PC-controllable	Label	Accessories/ connectivity	Size	References
FS-Scanner	Maxwell Sensors	Yes	Yes	Yes	Yes	• Fluorescence	• USB to PC • Software • Customization	Desktop	[24]
Uniscan	Unison Bioscience	No	Yes	No	Yes	• Colorimetric only	• PC connection • Printer • Software	Handheld	[25]
Skannex Document Scanner	Skannex	Yes	Yes	Yes	No	• Colorimetric only	• PC connection • Software	Desktop	[26]

Several entries can neither be grouped into handheld nor portable readers.
*OEM: original equipment manufacturer
N/A°: not applicable or no information available

focuses on handheld and portable readers, a good overview of larger commercially available readers is listed here as well.

9.4 Factors Compromising Accurate Quantitative Results

9.4.1 Positioning Errors

Disposable readers, such as the digital pregnancy test reader or the Alverix reader, are usually designed to provide qualitative yes/no answers. To obtain quantitative results, certain factors and design issues have to be considered to assure an accurate test outcome. The first positioning error is the relative positions of the test and control bands during printing of capture molecules onto the strip membrane (see Chapter 1). These bands can be closer or farther away from each other or be shifted relative to the strip ends. The second positioning error is cutting of the membranes, resulting in slightly different lengths. The third positioning error is the placing of the membrane strip into the plastic cassette. The strips can be positioned a little bit to the right or to the left relative to the reading window and the cassette housing. All three errors can add up to large errors. The following consideration highlights the influence on the result of a quantitative measurement when positioning errors occur. Figure 9.1

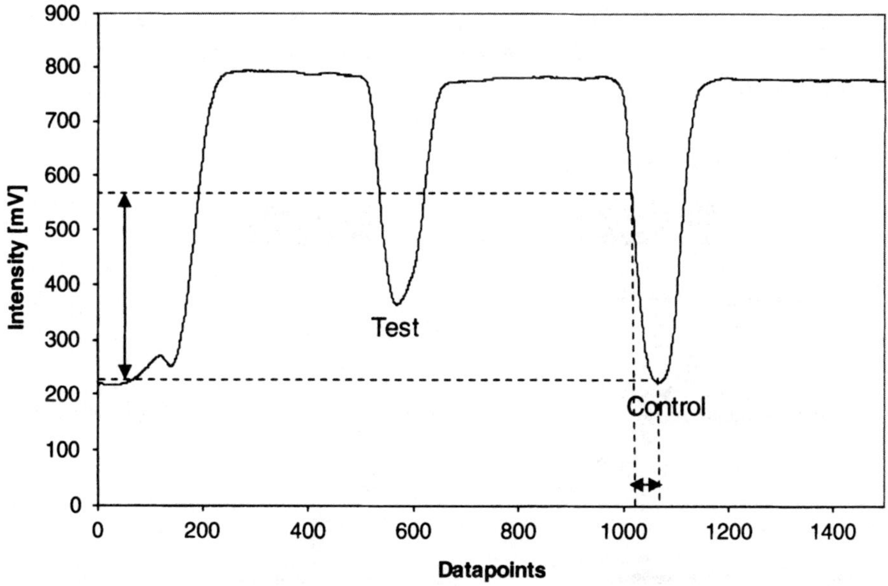

Fig. 9.1 Intensity profile of a test and control band in a reflectrometric measurement
A shift in horizontal positioning of the test strip (=positioning error) of 4% can half the intensity measured when scanning is not used.

shows a typical test and control band intensity profile. A shift in positioning of the control band for 4% results in a signal decrease of about 50% of the intensity measured. If a test is measured using a stationary detector, which is looking at a particular point on the strip (off-axis system, see below), then a positioning error of the strip can give erroneous results and lead to a completely wrong diagnostic conclusion and a misleading recommendation for a subsequent therapeutic procedure.

9.4.2 Off-Axis Versus Confocal Measurements

Not only is the positioning of the strip membrane in the horizontal orientation important but also its positioning in the vertical dimension is as important. The vertical dimension is the distance between the detector and the sample strip. For any optical measurement system, the sensitivity of detection changes with distance of the sample to the detector and is usually expressed as numerical aperture. This factor is particularly critical for a commonly used illumination-detection configuration called "off-axis" in which changes in distance may result in a total loss of signal. A solution to this problem is provided by a different design known as "confocal" detection. Both configurations are illustrated in Fig. 9.2. The sample at position A is to initiate signal transmission in both configuration designs. However, a shift of the sample to position B results in a signal for the confocal detectors but not for the off-axis detectors. In fact, the off-axis geometry looses the signal completely because the detector is focused on a spot that is not illuminated, and which does not contain the test or control line.

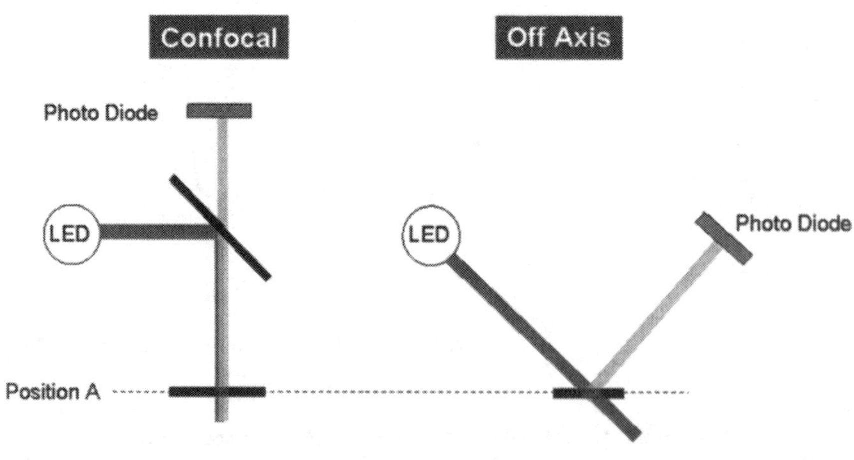

Fig. 9.2 Confocal versus off-axis principles and its influence on sample positioning

9.4.3 Accurate, Sufficient, and Homogeneous Illumination of the Sample

If the test strip is not illuminated properly, the test and control bands on the test strip will not be read correctly by the detector, irrespective of whether a stationary, a scanning, or a CCD/CMOS system is used. This factor is usually greatly underestimated. For homogeneous illumination of a sample, typically the illuminated area must be a multiple of the sample size unless very special techniques are used. A system that illuminates a larger area, such as an entire lateral flow strip at a time, must provide homogeneous illumination over the entire strip for meaningful results. The intensity of illumination is also very important to achieve the best sensitivity. This requires a large light power. The sensitivity of optical systems is further influenced by the distance of the detector to the sample, a factor that is known as numerical aperture.

9.4.4 Numerical Aperture, Field of View, and Sensitivity

The closer the detector is positioned to the illuminated sample, the more photons it can capture. Therefore, it is important to position the detector close to the sample. However, systems that need to illuminate the entire strip area cannot be positioned at any desired distance to the sample as the field of view limits the positioning. If a CCD camera is positioned very close to the sample, it cannot view the entire strip and the images at the edges get distorted due to the wide angles. The wider this angle the larger are lens distortion effects, which need to be compensated. A confocal scanning system, on the other hand, can be positioned very close to the sample and since it scans, it does not need to view the entire strip at a time. This is of particular importance when multiple test lines need to be read with equal accuracy on a single test strip.

9.4.5 Calibration Curve, Analyte Concentration, and Dynamic Range

The shape of the calibration curve that determines the concentration of the analyte also dictates the degree of accuracy of the results. For example, an S-shaped calibration curve of a test allows more accurate results when the analyte concentration being measured falls within the steepest (linear) range of the curve rather than into the sections where the slope is smaller (Fig. 9.3). At the asymptotic sections of the curve, no determination of the analyte concentration is possible at all. However, a qualitative result can often be derived using cut-off levels (>, <).

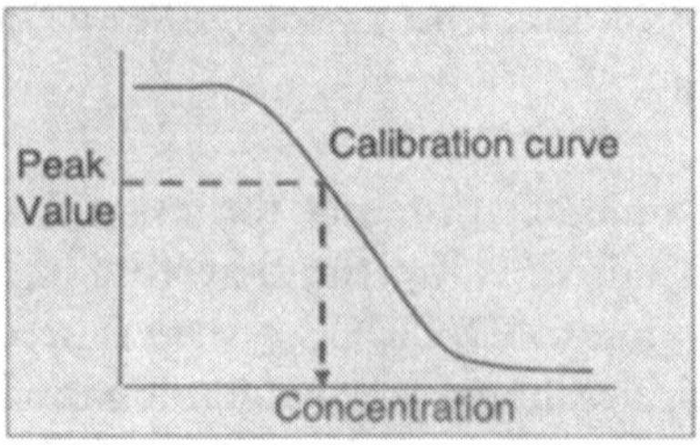

Fig. 9.3 Relationship of analyte concentrations and shape of calibration curve
Analyte concentration can most accurately be determined in the linear section of the curve. This is where the slope of the standard curve is highest.

9.4.6 Other Factors

There may be other factors influencing accurate quantitation of samples such as variability of sample composition, appearance, influences of drugs when samples are drawn from individual patients undergoing therapy, and others more. Those are not subject to this paper and will not be discussed here.

9.5 Portable Lateral Flow Scanner Readers for Immunoassays Based on Gold and Fluorescence Labels

9.5.1 Design Principles

In a portable scanning reader, colorimetric scanning of the lateral flow test strips is performed by moving the test strip over a light source or vice versa (Fig. 9.4). The instrument is adjusted so that light reflected from the strip membrane is received by a confocal detector and registered as a high signal. When the control band passes the light beam, the light reflected is decreased because the colloidal gold particles in the band absorb light at that wavelength. This reduced light intensity is shown as a negative band. Gold particles used in lateral flow techniques are typically 40 nm in size and have an absorption maximum at 540 nm. Absorption and reflection of the light are proportional to the overall density of colloidal gold particles of the band. Therefore, accurate quantification of colorimetric lateral flow tests is easily achieved. It should be pointed out that the measurements described here are not absorption measurements as light source and detector are arranged in a confocal (0° angle) and not a 180° configuration. The measurements may be best described as reflectance measurements.

For fluorescence measurements, the scanning process is the same as that described above for colorimetric assays. The only difference is that the reader now contains a fluorescence rather than a reflectrometric sensor. The sensor moves across the strip and excites the fluorescently labeled control and test

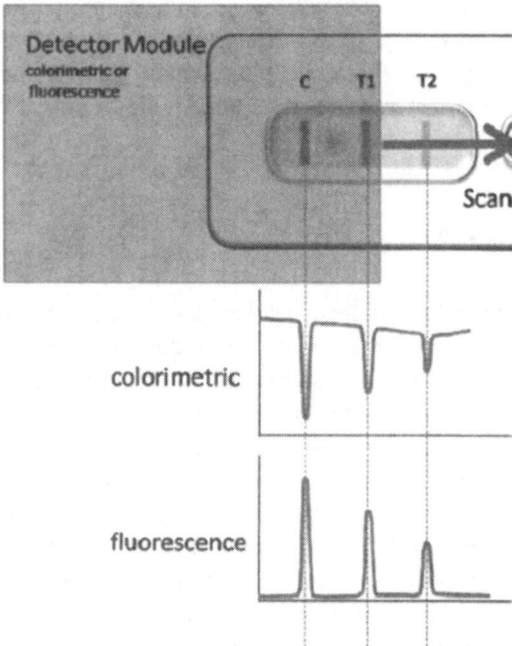

Fig. 9.4 Design principles of a detector module for colorimetric or fluorescence measurements A projected light dot is moved across the lateral flow strip. In reflectrometric measurements, the light is reflected back into the reader. At the peaks, less light is reflected due to absorption of the light by the label. Therefore, peaks for reflectrometric (colorimetric) measurements are below the baseline (negative peaks). The label can be of any color as the reader can be configured with various wavelengths.

lines. The fluorescent light of the test and control lines, which comes back into the confocal reader, results in an increase in fluorescence in the sensor and therefore positive peaks (above the baseline) are recorded (Fig. 9.4).

Fluorescence labels are used today in almost any industry and research area. By replacing colloid gold particles with fluorescence labels in lateral flow immunoassays, a more sensitive readout can be achieved. One example of a fluorescence lateral flow assay technique called Fluorescence Labeled Optical-Read Immune Dipstick Assay (FLORIDA) has been described by Bonenberger et al. [2] (see Chapter 1). Sensitivity of the assay was increased by 100- to 1,000-fold. This catapults lateral flow immunoassays into a sensitivity range of parts per trillion (ppt) and adds a new dimension to the technique as it can now be applied to new areas and markets.

It should be pointed out that the dissociation constant (K_d) values of the antibodies are also important factors, which influence sensitivity. If the K_d are in the nano-molar range, pico-molar sensitivities cannot be achieved, no matter which label is used.

Fluorescence can be read in a variety of ways, including intensity measurements, time-resolved fluorescence, phase modulation, two-photon excitation, and fluorescence anisotropy. Of those, simple intensity measurements are the cheapest and easiest to perform and are most applicable to lateral flow immunoassay test strips.

9.5.2 The ESE-Quant Reader

An example of a portable lateral flow reader that works according to the principles outlined above is the "ESE-Quant" manufactured by ESEGmbH of Stockach, Germany [8] (Fig. 9.5a). The diameter of the light spot projected onto the surface of the strip is about 1.1 mm. The distance of the light source to the strip surface is about 6 mm, which allows for a good numerical aperture. A simple LED serves as the light source. This, together with the miniaturized optoelectronic core [5, 6], provides a cost-efficient reader device. Data acquisition rate is up to 1,500 data points per second. One scan takes only a couple of seconds. The stand-alone reader can automatically find the peaks of the control band and up to 16 test band(s) and stores the data in the internal memory, without the need of connecting to an external computer. The ESE-Quant reader displays the results on an internal display and can be connected via an USB port to any computer. It has further connectivity (Fig. 9.5b) to a barcode reader, a printer, an RFID management system, and a charger to recharge the internal batteries. Data storage of the full measurement curves (raw data), peak intensities and areas, position of the peaks, ratios of the peaks, applying various standard curves and calculating analyte concentrations, background/baseline correction, as well as many further data analysis can be performed. Wireless data transfer via BlueTooth or GSM is an optional add-on as is a reference material that can serve as an internal standard.

An overview of the process from data recording to analyzed quantitative result is shown in Fig. 9.6a. Briefly, the peak intensities or peak areas are derived from the raw data curve including baseline and background correction. Standard curves are then applied to those values and the results are converted into an analyte concentration. Baseline and background correction are shown

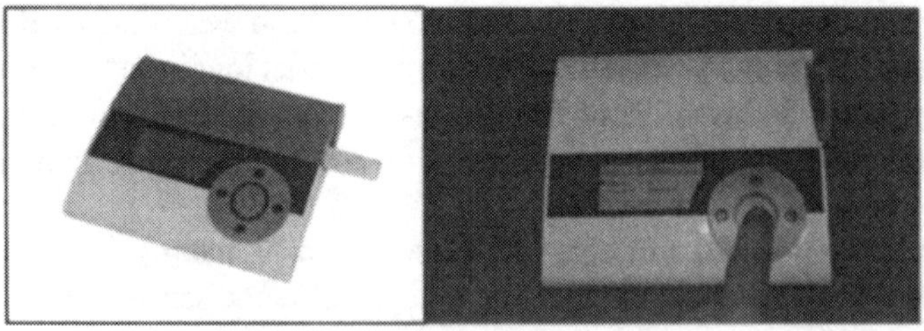

Fig. 9.5a Two models of the ESE-Quant Lateral Flow Reader (Size ca. 15 × 20 × 5 cm W × L × H)

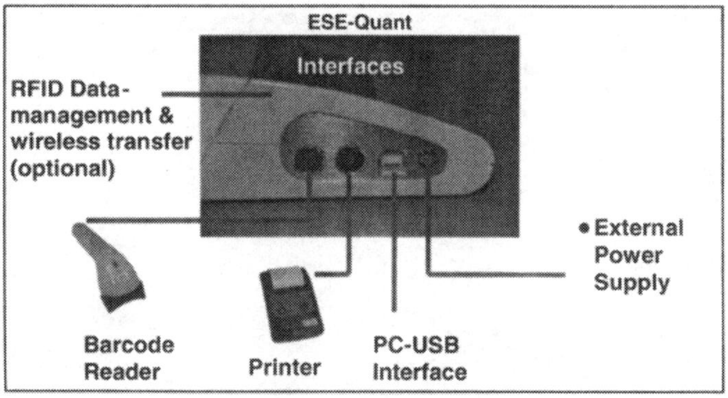

Fig. 9.5b Connectivity of the stand-alone ESE-Quant reader

in more detail in Fig. 9.6b. The baseline is extrapolated and the peak height or peak area calculated based on the new baseline. The manufacturer of the tests has the opportunity to configure the reader for different tests with parameters that account for the appropriate baseline correction. Different peaks can be corrected and evaluated independently using different parameters and different standard curves applied to different test lines.

In cases where the standard curves do not allow for accurate quantitative measurements, qualitative cut-off values can be applied to yield a positive/negative or high/low or yes/no answer (e.g., Fig. 9.3). The reader recognizes automatically whether a test is invalid. This is the case, for example, when no control line is found, when the intensity of the control line is too low, or if the test and

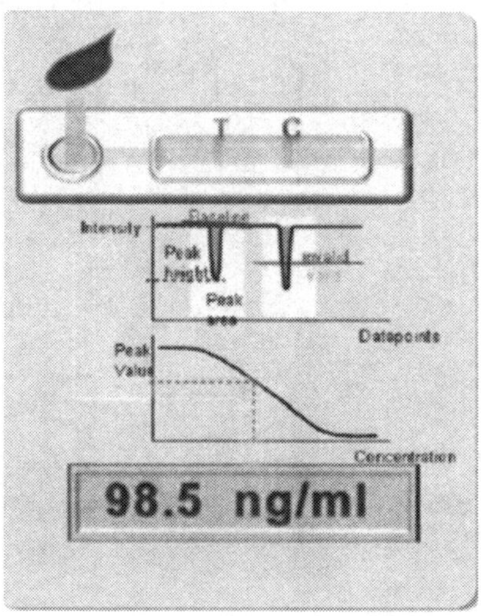

Fig. 9.6a From scan to analyzed results: an overview

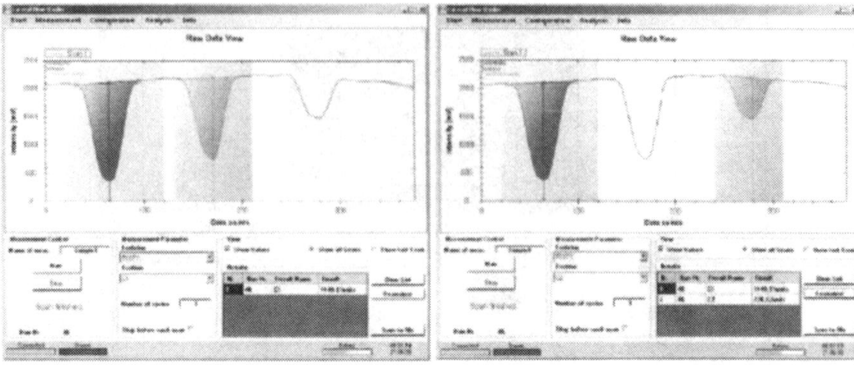

Fig. 9.6b Independent evaluation of different test lines and baseline correction

control lines do not appear in the expected range. Finally, the results are displayed on the reader display and in a downloadable results table (Figs. 9.6c and 9.6d), which contains all information about the test, the reader, the time and date of measurement, operator ID, patient information (if entered), and an automatically assigned run number for backtracking purposes. Results are write-protected. It is emphasized that the results table, the workflow of operation, and data information can easily be configured with the self-explanatory software and firmware provided. Complete customization is also always possible.

The ESE-Quant reader can be customized to read any format of lateral flow cassettes or just strips from any test manufacturer, and be configured to read up

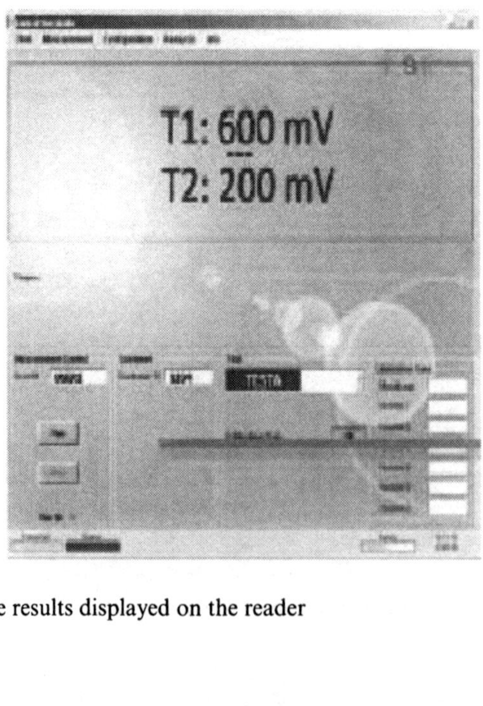

Fig. 9.6c The test line results displayed on the reader

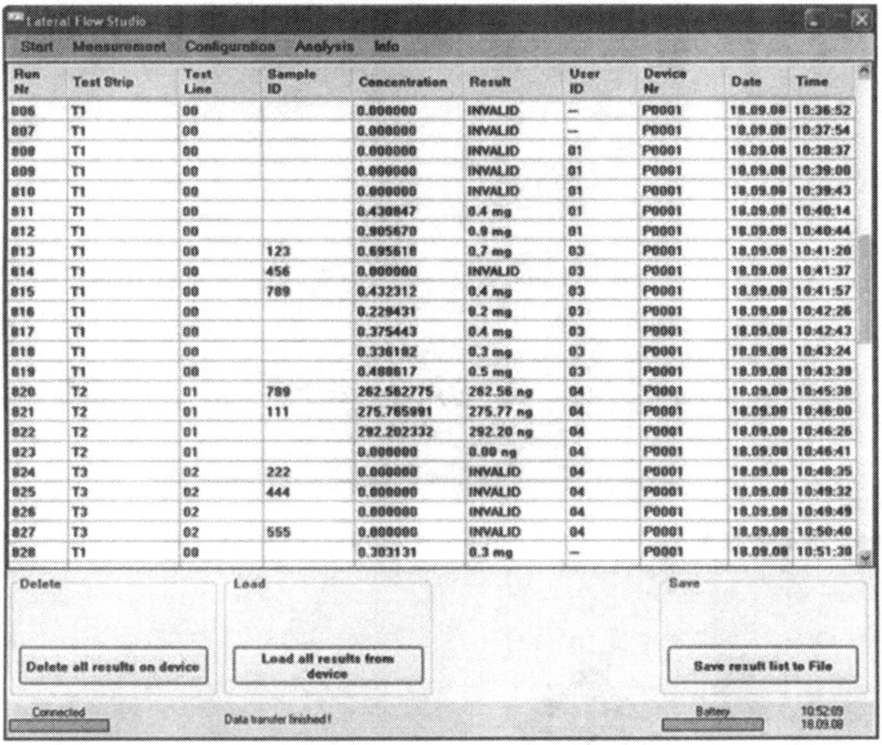

Run Nr	Test Strip	Test Line	Sample ID	Concentration	Result	User ID	Device Nr	Date	Time
806	T1	00		0.000000	INVALID	--	P0001	18.09.08	10:36:52
807	T1	00		0.000000	INVALID	--	P0001	18.09.08	10:37:54
808	T1	00		0.000000	INVALID	01	P0001	18.09.08	10:38:37
809	T1	00		0.000000	INVALID	01	P0001	18.09.08	10:39:00
810	T1	00		0.000000	INVALID	01	P0001	18.09.08	10:39:43
811	T1	00		0.430047	0.4 mg	01	P0001	18.09.08	10:40:14
812	T1	00		0.905670	0.9 mg	01	P0001	18.09.08	10:40:44
813	T1	00	123	0.695618	0.7 mg	03	P0001	18.09.08	10:41:20
814	T1	00	456	0.000000	INVALID	03	P0001	18.09.08	10:41:37
815	T1	00	789	0.432312	0.4 mg	03	P0001	18.09.08	10:41:57
816	T1	00		0.229431	0.2 mg	03	P0001	18.09.08	10:42:26
817	T1	00		0.375443	0.4 mg	03	P0001	18.09.08	10:42:43
818	T1	00		0.336182	0.3 mg	03	P0001	18.09.08	10:43:24
819	T1	00		0.488817	0.5 mg	03	P0001	18.09.08	10:43:38
820	T2	01	789	262.582775	262.58 ng	04	P0001	18.09.08	10:45:38
821	T2	01	111	275.765991	275.77 ng	04	P0001	18.09.08	10:46:00
822	T2	01		292.202332	292.20 ng	04	P0001	18.09.08	10:46:26
823	T2	01		0.000000	0.00 ng	04	P0001	18.09.08	10:46:41
824	T3	02	222	0.000000	INVALID	04	P0001	18.09.08	10:48:35
825	T3	02	444	0.000000	INVALID	04	P0001	18.09.08	10:49:32
826	T3	02		0.000000	INVALID	04	P0001	18.09.08	10:49:49
827	T3	02	555	0.000000	INVALID	04	P0001	18.09.08	10:50:40
828	T1	00		0.303131	0.3 mg	--	P0001	18.09.08	10:51:38

Fig. 9.6d An example of a results table output

to 16 different tests and test lines. The reader can also be set up for any and multiple labels, be it a colored label (colloidal gold beads or any other color) or fluorescence label (from 365 to 720 nm is standard). This is called "integrated flexibility". Customers appreciate the quick turnaround time, the low investment needed for the customization, and the low technical and financial risk. They can approach the market quickly with a proven lateral flow reader technology as they do not need to wait and pay for the development of a reader from scratch. Users can also maintain their cassette configuration and do not need to change their tests or products. ESE GmbH has been the first company to provide this degree of integrated flexibility on the market.

9.6 Handheld Lateral Flow Scanner Readers for Immunoassays Based on Gold and Fluorescence Labels

Besides portable readers, ESE has also developed handheld reader device for lateral flow immunoassays both for colloidal gold-based colorimetric readout (*ESE GoldScan*) and fluorescence readouts (*ESE FluoScan*) [7] (Figs. 9.7a and 9.7b) to assist the operator with high performance and convenience. The unique

Fig. 9.7a Battery-operated handheld lateral flow reader prototypes by ESE for colorimetric readout using colloidal gold beads or color latex beads (*GoldScan*) and fluorescence labels (*FluoScan*)

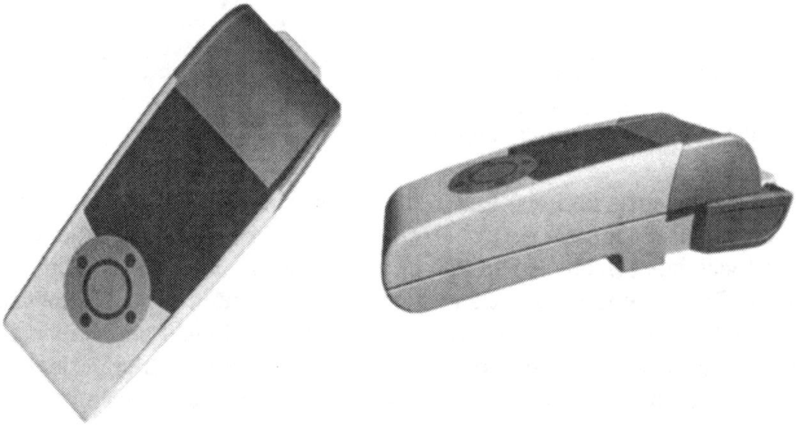

Fig. 9.7b Two next-generation serial handheld lateral flow readers by ESE, both for gold- and fluorescence-based lateral flow assays

fluorescence reader technology makes the new generation of immunoassays usable in many new applications. With the latest micro optics technology and highly integrated electronics, the sensitivity of the fluorescence sensor is comparable to expensive benchtop instruments. The reader is operated by inserting the test strip and pressing the start button. The device automatically scans, calculates the results, and documents the measurements. Up to 2,000 scans can be stored in the device. Transfer of the results via USB provides complete PC functionality for further documentation, printing, and data storage. Wireless communication is available as an option. The device can be customized for specific fluorescence dyes from 365 to 720 nm and for various cassette formats. It covers all types of market applications of lateral flow tests.

9.6.1 Characterization of the Reader Devices

ESE lateral flow readers are currently marketed by several tests manufacturers for various applications such as point-of-care testing, physician's office laboratory (POL) drug testing, allergy testing, infectious diseases, environmental pollutants, toxins in food and agricultural products field based testing and others more.

Figure 9.8 shows a representative scan of a colorimetric lateral flow immunoassay test strip using 40-nm gold beads as labels. About 50 μl urine sample spiked with various concentrations of tetrahydrocannabinol (THC) were applied to the sample port of the strip in a commercially available drug-of-abuse test cassette. The assay tests for THC drug consumption in blood and urine. The scan was recorded after the sample was applied for 9 min. One thousand five hundred data points were recorded in 1 second. The control and test bands are clearly visible, and the intensity is dependent on the concentration of the analytes. Results indicate that there is a good correlation between signal intensities and drug concentrations. Peak positions of the control

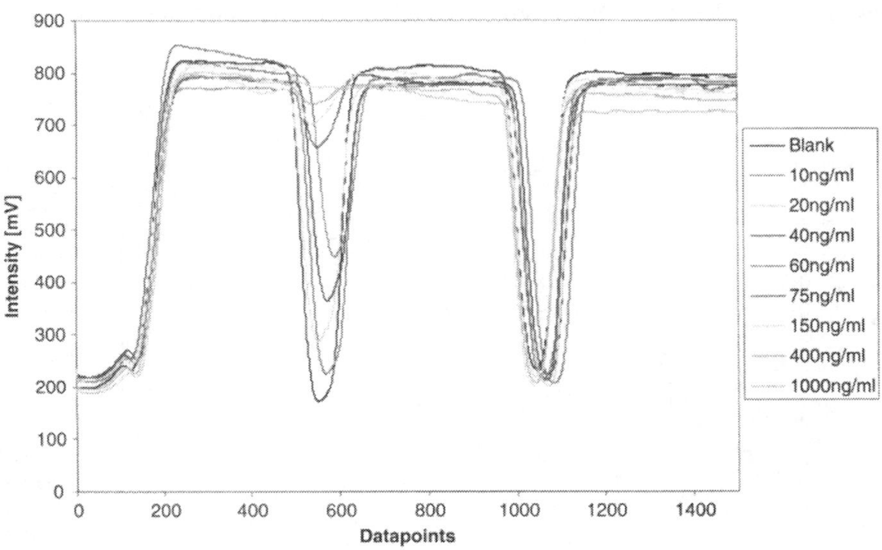

Fig. 9.8 Raw data of lateral flow immunoassays using gold beads (*ESE GoldScan*), and different concentrations of THC spiked into urine as samples
Test and control bands are visible. At the very left (0–200 data points), part of the cassette housing is visible.

Table 9.3 Reproducibility of peak positions of the nine different strips from Fig. 9.8

Conc. [ng/ml]	0	10	20	40	60	75	150	400	1000
Control peak position	549	567	548	569	581	550	541	542	556
Test peak position	1043	1057	1040	1069	1081	1065	1038	1040	1072
Difference in position T-C	494	490	492	500	500	515	497	498	516
	Average	Std dev.	%Std dev.						
Control band	556	13.62	2.45						
Test band	1056	16.34	1.55						
Difference in position T-C	500	9.31	1.86						

bands and test bands on the nine scan all fall within a narrow range. The % standard deviation is 1.55–2.45. These results are summarized in Table 9.3.

The variations seen in the last experiment are mainly due to biochemical parameters as can be seen from the following reproducibility studies of the reader. In this experiment, a urine sample was spiked with 25 ng per ml of THC and applied to the lateral flow immunoassay cassette. The same cassette was scanned 20 times using the *GoldScan* handheld device. Results (Fig. 9.9) show that reproducibility was excellent. Again Tables 9.4 and 9.5 summarize the values of the peak intensity and peak positions of the 20 control bands and test bands on the scan. The standard deviation is in the range of 0.25–0.40%.

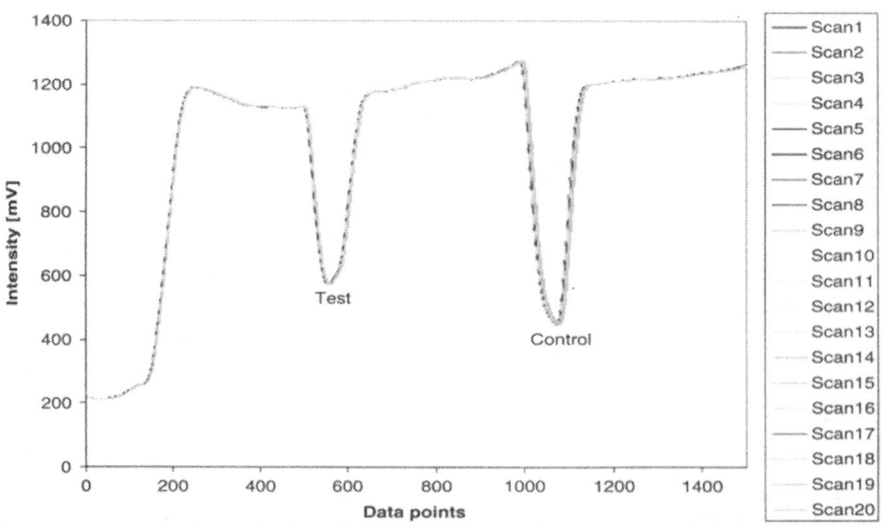

Fig. 9.9 Reproducibility study: a urine sample spiked with 25 ng per ml THC was applied to the lateral flow immunoassay and scanned 20 times using the *GoldScan* handheld device

Table 9.4 Peak intensity and standard deviation data from experiment shown in Fig. 9.9

	Scan 1	Scan 2	Scan 3	Scan 4	Scan 5
Test [mV]	572.75	572.87	572.71	572.56	572.71
Control [mV]	448.89	448.84	449.05	448.97	449.27
Ratio C/T	0.784	0.783	0.784	0.784	0.784
	Scan 6	Scan 7	Scan 8	Scan 9	Scan 10
Test [mV]	572.85	572.88	572.61	572.63	572.32
Control [mV]	449.17	448.97	448.86	448.90	448.93
Ratio C/T	0.784	0.784	0.784	0.784	0.784
	Scan 11	Scan 12	Scan 13	Scan 14	Scan 15
Test [mV]	572.82	573.03	572.66	572.75	572.82
Control [mV]	449.32	449.23	449.21	449.49	449.05
Ratio C/T	0.78	0.78	0.78	0.78	0.78
	Scan 16	Scan 17	Scan 18	Scan 19	Scan 20
Test [mV]	572.85	573.08	572.86	572.74	573.47
Control [mV]	448.81	449.14	449.15	448.94	449.17
Ratio C/T	0.783	0.784	0.784	0.784	0.783
	Average	Standard dev.	%Std dev.		
Testband	572.80	0.22	0.039		
Controlband	449.07	0.18	0.040		
Ratio C/T	0.784	0.004	0.047		

Table 9.5 Peak position and standard deviation data from experiment shown in Fig. 9.9

	Scan1	Scan2	Scan3	Scan4	Scan5
Test peak position	556	554	554	554	556
Control peak position	1073	1069	1070	1070	1073
Difference C-T	517	515	516	516	517
	Scan6	Scan7	Scan8	Scan9	Scan10
Test peak position	555	557	556	556	559
Control peak position	1073	1074	1073	1074	1077
Difference C-T	518	517	517	518	518
	Scan11	Scan12	Scan13	Scan14	Scan15
Test peak position	558	559	557	558	559
Control peak position	1076	1078	1074	1076	1077
Difference C-T	518	519	517	518	518
	Scan16	Scan17	Scan18	Scan19	Scan20
Test peak position	561	560	559	561	560
Control peak position	1078	1079	1078	1078	1081
Difference C-T	517	519	519	517	521
	Average	STD DEV	%STD DEV		
Testband	557	2.22	0.40		
Controlband	1075	3.19	0.30		
Difference C-T	518	1.28	0.25		

The *ESE-Quant* and *FluoScan* lateral flow readers are also appropriate for scanning fluorescence latex beads. An example is shown below for drug screening, but the devices can be applied to any fluorescence lateral flow tests. In an experiment performed by Diagnostic Consulting Network (DCN, Carlsbad CA), MDMA (Ecstasy) was spiked into urine samples at different concentrations (0, 125, 150, 250, 500, 750, and 1,000 ng per ml) to obtain a standard curve for quantification purposes. A competitive assay format was chosen. Four test strips for each concentration were used. About 150 µl of spiked sample was applied to the test strip and the strips scanned after 15 min using the ESE *FluoScan* handheld device. Figure 9.10 shows the results of this experiment. The standard curve could be fitted by linear regression. The correlation coefficient is $R^2 = 0.9958$, indicating an excellent fit of data.

Quantification of results can be performed by using a standard curve as external standard (Fig. 9.10), but an internal standard can also be used. ESE has developed stable reference materials, which can be implemented into the readers for even more precise readout. In addition, the control band can also be used as internal standard if a reproducible intensity is obtained. These are relative measurements and the ratio of the intensities of the internal standard to the test band can often give a more precise data evaluation.

9.6.2 Fluorescence Visualizers

For customers who want to take advantage of the highly sensitive fluorescence technology and require only visual read-out, ESE has developed another

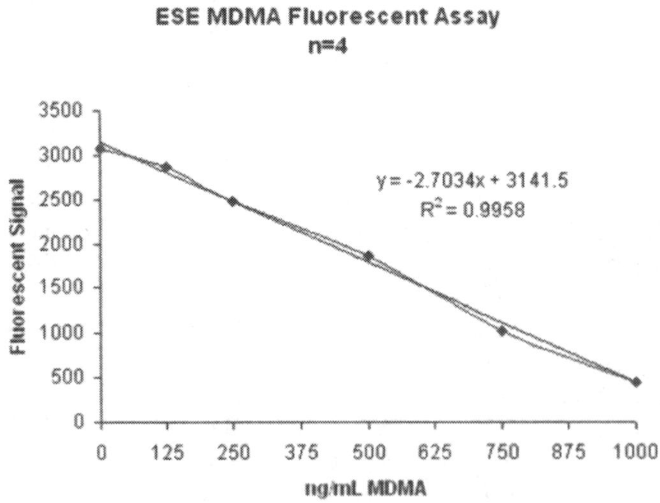

Fig. 9.10 Standard curve of MDMA-spiked urine samples. Strips were scanned using the *FluoScan* handheld fluorescence lateral flow reader

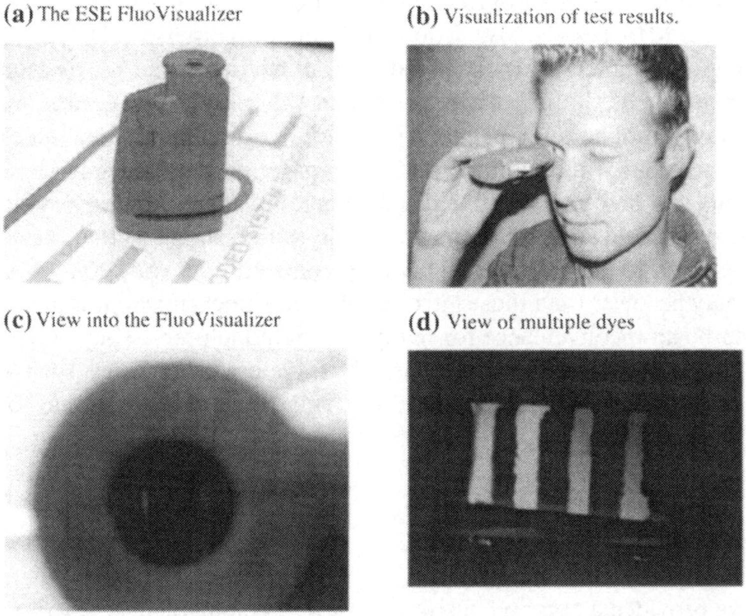

Fig. 9.11 A handheld, pocket-sized visualizer for fluorescence-based lateral flow test strips

simpler and cheaper reader for this purpose. The FluoVisualizer (Fig. 9.11a) is a handheld, pocket-size device for lateral flow immunoassays using fluorescence dyes [7]. The reader contains an illumination source and a filter. The human eyes can visualize the test results. No electronic recording of data is performed. Since fluorescence is much more sensitive, it allows for higher multiplexing capacity. The reader is simple in design and therefore very affordable. Figure 9.11b shows the operation of the device. In these procedures, the test strip is inserted, a button pressed to turn on the excitation source, and the operator looks inside the device to see whether control and test bands are visible. Figure 9.11c is a photograph of the results shown in the direction of view of the operator. A test band and a control band are visible. Figure 9.11d shows that the visualizer can detect multiple dyes at a time and is therefore suitable for multiplexing. Dyes from Sensopath [27] were used for this study.

9.7 Reader Devices for Magnetic Particle–Based Assays

Readers for magnetic particle–based lateral flow immunassays are similar to those described for scanning systems. A good overview of this technology is provided by LaBorde et al. [28]. Commercial reader devices are developed by British Biocell International (BB International, Cardiff, United Kingdom) [29]. A handheld unit, *MICT MAR*^{TM} (magnetic assay reader), is available from

Magna Biosciences/Quantum Design (San Diego, CA, USA) [30]. A tabletop reader, the MIATek, is also available from Magnisense [21]. The biggest advantage of magnetic particle–based readout devices is that the measurement is independent of the impurities present in the samples. Biological matrices commonly do not contain magnetic particles, resulting in minimal interference of assay sensitivity and minimal sample preparation requirement. These measurements can be quite sensitive disadvantages of these devices, however, are that magnetic field strength decreases rapidly with distance so that the detector must be close to the particles. This may contribute to variations in signals, which may be lower than those introduced by the biochemical procedures. This can also be an inconvenience for many test manufacturers. They may have to modify their devices as most established assays use other labels such as gold beads or fluorescence labels and need to change their cassette format to fit the test into the reader.

9.8 Summary

We have provided a comprehensive overview of commercially available lateral flow immunoassay readers, with strong emphasis on handheld and portable devices. We have pointed out trends in the industry and addressed detection issues pertinent to different systems. Examples and characterization of some readers are discussed. Different detection schemes are compared and advantages and limitations highlighted. We have covered colorimetric, fluorescence, and magnetic readout technologies. Additionally, we have shown that lateral flow immunoassays can now be performed quantitatively, accurately, reproducibly, and with increased sensitivity, enabling new applications of the lateral flow immunoassays.

Acknowledgments We would like to thank Dr. Brendan O'Farrell of Diagnostic Consulting Network (DCN) and Dr. Bernhard Gerstenecker of Milenia Biotec for helpful discussions and evaluation of lateral flow immunoassay readers.

References

1. O'Farrell, B. and Bauer J. (2006) Developing highly sensitive, more-reproducible lateral-flow assays. Part 2: New challenges with approaches. IVD Technology 7:67.
2. Bonenberger, J. and Doumanas, M. (2006) Overcoming sensitivity limitations of lateral-flow with a novel labeling technique. IVD Technology 5:41–46.
3. Piepenburg, O., Williams, C. H., Stemple, D. L. and Armes, N. A. (2006) DNA detection using recombination proteins. PLOS Biol. 4:e204,1–7.
4. Seal, J., Braven, H. and Wallace, P. (2006) Point-of-care nucleic acid lateral-flow tests. IVD Technology 8:41–54.
5. Faulstich, K., Gruler, R., Eberhard, M. and Haberstroh, K. (2007) Developing rapid mobile POC systems. Part 1: Devices and applications for lateral-flow immunodiagnostics. IVD Technology 13(6):47–53.

6. Faulstich, K., Gruler, R., Eberhard, M. and Haberstroh, K. (2007) Developing rapid mobile POC systems. Part 2: Nucleic acid based testing platforms. IVD Technology 13(7):47.
7. www.ese-gmbh.de
8. European Patent Office Publication Number 0088636, 14.09.83, Bulletin 83/37.
9. www.biosite.com
10. www.cholestech.com
11. www.unipath.com
12. www.cozartgroup.com
13. www.alverix.com
14. www.responsebio.com
15. www.charm.com
16. www.alexeter.com
17. www.neogen.com
18. www.matest.de
19. www.e77.hu
20. www.qdusa.com
21. Lenglet, L., Nikitin, P. and Pequignot, C. (2008) Magnetic immunoassays: a new paradigm in POCT. IVD Technology July/August, pp. 43–49.
22. Müller-Bardorff, M., Sylvén, C., Gundars, R., Jørgensen, B., Collinson, P. O., Waldenhofer, U., et al. (2000) Evaluation of a point-of-care system for quanti-tative determination of troponin T and myoglobin. Clin. Chem. Lab. Med. (2000) 38:567–574.
23. http://jp.hamamatsu.com/products/life-science/3005/C10066/index_en.html
24. http://www.maxwellsensors.com/FS-Scanner_Flyer.pdf
25. http://www.tubi.com.tw/Uniscan_rapid%20test%20reader.htm
26. http://www.skannex.com/how.html
27. www.sensopath.com
28. LaBorde, R. T. and O'Farrell, B. (2002) Paramagnetic-particle detection in lateral-flow assays. IVD Technology 4:36–42.
29. http://www.britishbiocell.com/magnetics.htm
30. http://www.qdusa.com/biotech06/mbs_tech.html

Chapter 10
Quantitative, False Positive, and False Negative Issues for Lateral Flow Immunoassays as Exemplified by Onsite Drug Screens

Raphael C. Wong and Harley Y. Tse

10.1 Introduction

Lateral flow immunoassay devices offer many advantages including convenience, economical, simplicity, and rapid result. Many lateral flow immunoassays are non-instrumental and rely on visual detection of colored lines for results, enabling easy portability and allowing testing at any time and at any place by non-technical personnel. Hence, many lateral flow immunoassay tests have been developed for use “onsite”, “point-of-care”, or “point-of-test”. However, utilization of this type of device requires the acceptance of some trade-offs. The most important is that the test results are generally qualitative. Moreover, since it is antibody-based, possibility exists that chemicals with similar structures will cause positive result leading to specificity and sensitivity issues. Recognizing such limitations, this technology has been widely utilized for screening tests in which a yes and no answer is sufficient. The initial screen test result can then be confirmed by a quantitative method, which is usually equipment-based and required the service of highly trained technician. Acceptance of this two-step procedure allows the massive testing of subjects economically and quickly without worrying about the test locations and the service of highly trained technician. An example of one such application is at home pregnancy testing in which a positive result would inevitably lead to a doctor’s visit for further testing. Another example is workplace drug testing in which positive result from a drug screen is confirmed by a laboratory GC/MS or LC/MS result.

With the increasing popularity of lateral flow immunoassay devices, many users have ignored the inherent qualitative nature and specificity/sensitivity issues of the assays and placed increasing demands on the manufacturers for fool-proof onsite tests. The present chapter examines this issue as exemplified by the false results encountered in abused drug screens. We shall first examine the definitions of positive and negative results in a

R.C. Wong (✉)
Branan Medical Corporation, 140, Technology Drive, Suite 400, Irvine, CA, USA
e-mail: raphael@brananmedical.com

R.C. Wong, H.Y. Tse (eds.), *Lateral Flow Immunoassay*,
DOI 10.1007/978-1-59745-240-3_10,

cut-off type test device, then present the generally accepted statistical analysis of such test results and finally investigate the various factors that may give rise to false results. Efforts to correct and improve the lateral flow immunoassay are also presented.

10.2 Background Information on Drug Screen

In the United States, SAMHSA (Substance Abuse and Mental Health Services Administration of the Department of Health and Human Services) publishes comprehensive drug testing guidelines including the cut-off concentration level for both the initial screen test and the GC/MS confirmation [1]. It is notable that for many drugs, the cut-off concentrations for GC/MS are lower than those for the drug screens. This is to accommodate the less than perfect specific nature of immunoassay in which the antibody may recognize both the target drug and its metabolites while the GC/MS is specific for either the specific target drug or the metabolite.

10.3 Statistical Analysis of Lateral Flow Immunoassay Results

A positive drug screen result means that the target drug is present in the specimen at a concentration above the cut-off level. The result does not indicate:

a) How the drug enters the test subject's body.
b) Whether the subject is a chronic user or a recreational user.
c) Whether the subject is under the influence of the drug when the test was administered.
d) How high above the cut-off level is the target drug present in the specimen.

A negative drug screen result shows that either there are no target drugs (or target drug metabolites) in the specimen or, if present, their concentrations are below the cut-off levels. However, it does not indicate if the subject took drugs sometimes ago and now the drugs have cleared out the body.

A false positive (FP) occurs when the test result indicates the presence of drug in the specimen with concentration equal to or above the cut-off concentration when actually there is no drug present or, if present, it is below the cut-off concentration (usually as determined by GC/MS). A false negative (FN) occurs when the test indicates there is no drug present or the drug concentration is below the cut-off concentration when actually it has been determined by GC-MS to be above the cut-off concentration. A true positive (TP) was the correct indication by a device that the drug concentration is equal to or above the cut-off concentration. A true negative (TN) is the correct indication by the device that the drug concentration is below the cut-off concentration.

Evaluation on the performance of lateral flow immunoassay devices is usually based on several parameters including sensitivity, specificity, and efficiency. Calculations on the three parameters are briefly described below.

The term sensitivity in drug screen describes the device's ability to determine the presence of a drug. It is calculated by the formula:

$$\text{Sensitivity} = \frac{(\text{TP})}{(\text{TP} + \text{FN})}$$

Specificity describes the device's ability to determine the absence of a drug and is calculated by the formula:

$$\text{Specificity} = \frac{(\text{TN})}{(\text{TN} + \text{FP})}$$

The device's ability to correctly determine the presence or absence of a drug is known as efficiency and is calculated by:

$$\text{Efficiency} = \frac{(\text{TN} + \text{TP})}{(\text{TN} + \text{TP} + \text{FN} + \text{FP})}$$

In some instances where lateral flow immunoassays are used to predict a disease condition, two additional parameters will also be considered. The Positive Predictive Value (PPV) is the parameter to predict the probability that a test positive is a true positive. It is calculated as:

$$\text{PPV} = \frac{\text{TP}}{\text{TP} + \text{FP}}$$

The Negative Predictive Value (NPV), the parameter to predict the probability that a test negative is a true negative, is calculated as:

$$\text{NPV} = \frac{\text{TN}}{\text{TN} + \text{FN}}$$

Using some of these parameters, one can evaluate a lateral flow immunoassay device. An example is an article published in the year 2000 by Peace, Tarnai, and Polis in which four onsite drug screens were compared [2].

10.4 Interpretation of Cut-Off Value by Developers of Lateral Flow Immunoassay

Despite the connotation of the word "cut-off", it is impossible to develop an immunoassay test in which a specimen containing any drug amount below the cut-off will give negative result and as soon as the drug concentration in the specimen exceeds the cut-off level, the test will instantly become positive. Instead, developers have interpreted and optimized test devices such that when specimens with drugs at cut-off concentrations are tested, 50% of the test results would show positive and the other 50% would be negative. As the specimens contain increasing amounts of drugs, more of the test results would show positive so that when the sample concentrations reach 150% above the cut-off, most of the results should be positive. On the other hand, as the sample concentrations are decreasing from the cut-off concentrations, more and more negative results would be reported so that at 50% below the cut-off, almost all of the results would be negative. An example of such test design is shown in Table 10.1. This table is adapted from a report by D. Moody et. al. in which two lateral flow immunoassay drug test devices were evaluated [3]. Although each device was tested for multiple drugs, only the data for the cocaine metabolite benzylecogonine (BE) and the drug morphine (MOR) are being shown here. For each device, increasing specimen concentrations from the cut-off gave more positive results and decreasing concentrations gave more negative results. Statistical analysis on these data (shown on Table 10.2) suggested that the sensitivity, specificity, and efficiency of device B are better than device A. The conclusions were confirmed when clinical samples were evaluated by both devices (Table 10.3).

In a study comparing the performance of two point-of-care urine tricyclic antidepressant drug screens with quantitative serum chromatographic analysis, Melanson et al. used similar technique for evaluation [4] and lamented that these lateral flow immunoassay devices could not provide quantitative results. In the following sections, we shall examine some of the issues affecting the results and provide suggestions to improve the lateral flow immunoassay tests.

Table 10.1 An example of the test results of two devices (A and B) for the drugs Benzylecogonine (BE) and Morphine (MOR), with the number of positives ($n = 10$) in each concentration (Ref. 3)

% Cut-off	BE		MOR	
	A	B	A	B
0	0	0	0	0
25	0	0	0	0
50	4.5	1	1	0
75	5.5	5.5	4	4
125	8.5	10	6.5	10
150	10	10	8.5	10
175	9	10	8.5	10

Table 10.2 Statistical analysis of the results in Table 10.1

	BE		MOR	
	A	B	A	B
TP	27.5	30	23.5	30
FN	2.5	0	6.5	0
TN	30	33.5	35	36
FP	10	6.5	5	4
Sensitivity (%)	91.6	100	78.3	100
Specificity (%)	75	83.8	87.5	90
Efficiency (%)	82.1	90.7	83.6	94.3
PPV (%)	73.3	82.2	82.5	88.2
NPV (%)	92.3	100	84.3	100

Table 10.3 Clinical sample studies for BE and MOR (noted: an ambivant result is classified as negative in this analysis)

	BE		MOR	
	A	B	A	B
No. of samples	240	240	240	240
TP	48	48	47	47
FN	0	0	2	2
TN	192	192	190	191
FP	0	0	1	0
Sensitivity (%)	100	100	95.9	95.9
Specificity (%)	100	100	99.5	100
Efficiency (%)	100	100	98.8	99.2
PPV (%)	100	100	97.9	100
NPV (%)	100	100	99.0	99.0

10.5 Factors Affecting Test Results of Lateral Flow Immunoassay Devices

Many factors will contribute to the qualitative nature of the lateral flow immunoassay and the following are discussed: (1) manufacturing issue; (2) operator errors; (3) effects of food, supplements, and beverages; (4) environmental factors; (5) sample manipulation; and (6) cross-reactivity problem.

10.5.1 Manufacturing Issues

Some of the false results in lateral flow immunoassay are due to poor test design, bad parts, and flaws in the manufacturing process. These factors are briefly discussed in the following sections.

10.5.1.1 Antibodies

The foundation of immunoassay is the utilization of specific antibody to quickly capture the corresponding antigen. Earlier use of polyclonal antibodies provide easily obtainable antibodies, but have the limitations of high cross-reactivity. The utilization of monoclonal antibodies has substantially increased the specificity and sensitivity of the immunoassay. Detailed discussion on the use of antibody can be found in Chapter 4. However, cross-reactivity remains a continuing and major concern in the development of a lateral immunoassay device. Further discussion on the cross reactivity issue will be found in the later part of this chapter.

It is important to realize that in competitive type lateral flow immunoassay like the drug screen, the competing antigen conjugate plays as important a part in the test performance as the antibody. Moreover, a good antibody for heterogeneous immunoassay like the ELISA may not be usable for lateral flow immunoassay.

When there is a choice of antibody available, the antibody that offers a sharp drop off in test signal around the cut-off concentration will give the best test, since positive results will quickly be obtainable as soon as the specimen concentrations increase from the cut-off values and vice versa.

10.5.1.2 Adhesives

As described in Chapter 3 of this book, the components of the test strips (sample pad, conjugate pad, nitrocellulose membrane, and absorbent pad) are all connected to each other to enable continuous liquid flow in the lateral flow immunoassay test design. These components are held in place with adhesive bonded to a vinyl backing. Although it seems innocuous, the wrong adhesive would create tremendous problems on the test performance. Adhesives that are too fluidic may leak through the pores of the nitrocellulose membrane or the spaces in sparsely weaved sample pad and conjugate pad materials and interact with the samples and/or the conjugates, causing inconsistent and erratic results. Also, adhesives that lose the adhesive property quickly will delaminate the material components and would again lead to erratic results. Hence, it is very important to identify the right adhesive during the development phase for the manufacture of the lateral flow immunoassay.

10.5.1.3 Taping

The principle of the lateral flow immunoassay requires that the material components be connected to each other sequentially via overlapping of the first component to the subsequent component. The right amount of overlap will allow a uniform liquid front to form and flow through the sample pad, then the conjugate pad solubilizing the conjugate in an even manner, and move onto the membrane in one single liquid front. Advantage of such movement is that the test and control lines will be formed quickly and uniformly. Once they are formed, they will not change in color intensity. Also, the background color of the membrane will become clear soon after the passage of the conjugate liquid front, enabling easier detection of the lines on the membrane. False results may occur if the taping of the components is inconsistent, causing differences in component overlaps from one lateral flow strip to another. Manual taping is especially prone to produce such a problem. Utilization of a precise taping guide or an automatic taping system (see Chapter 8) can help to eliminate this problem.

10.5.1.4 Membrane Powder

A common practice to ensure overlap between the conjugate pad and the nitrocellulose membrane is the use of a thin cover tap (see Chapter 3) to hold the conjugate pad down onto the membrane. Such practice is especially popular with the manufacture of naked lateral flow immunoassay dipstick in which no plastic housing part is available to ensure the conjugate pad/membrane contact. However, quite a few nitrocellulose manufacturers supply membranes with powders on the surface (see Chapter 6) with the amount of powder varying among different membrane production lots. The existence of the powder will prevent the cover tap to adhere to the membrane, leading to varying degrees of overlap. This strip-to-strip variation causes inconsistent results. In case the conjugate pad fails to contact the membrane completely, then the test will fail with no liquid flow on the membrane and no control or test line will form.

To overcome this problem, assay manufacturers should pay great attention on the choice of the membrane supplier and be vigilant in monitoring the quality of incoming membrane material.

10.5.1.5 Plastic Housing

In most lateral flow immunoassay test devices, plastic housings are usually utilized for the following purposes:

a. To protect the test strip
b. To facilitate the handling of the test strip by the operator
c. For cosmetic reason
d. To incorporate other accessories (e.g., the drug test cup adds sample collection and shipping features in addition to testing purpose)

Most of these plastic housing are designed with certain features like pressure bars and sample orifice to help consistent liquid flow on the test strips. Since molded parts for a device are produced from several mold cavities, molded part from one cavity may deviate from another mold cavity, leading to inconsistent test result from one device to the next. Moreover, many housings are designed such that the strip is placed on a bottom plastic part and covered with another plastic top. If the closure of the top and bottom parts vary from device to device, then the liquid flow will vary too due to variation of pressure point. This again will lead to inconsistent result.

To avoid this situation, manufacturer needs to utilize a reputable mold maker and insertion molder to ensure that the molded parts are of consistent quality.

10.5.2 Operator Errors

Operator of the test is also a major source of erroneous test results. Most of the factors, however, can be easily corrected by better training. Some of the issues are briefly discussed below.

10.5.2.1 Flooding the Test

The sequential passage of liquid from one test component to the next in a lateral flow immunoassay test format is only possible if adequate, but not overly excessive, amount of sample liquid is used. Adding over-abundant amount of the liquid sample may cause portion of the liquid to bypass the sequential passage of components while the remainder portion is still moving from one component to the next. This will lead to "flooding of the test". In urine drug screen, this can happen when either a test dipcard is placed much further into the specimen container than designed or excessive amount of urine is added to the sample hole in the test cassette. Both these misapplications may cause urine sample to enter the membrane component, without passing the antibody–colloidal gold dye component. The end result is that test lines will not form. Or, if formed, they will appear in an undefined manner. Interpretation of these lines usually is erroneous.

This problem is easy to eliminate. Usually, it will go away once the operator receives adequate training.

10.5.2.2 Insufficient Sample Volume

The lateral flow immunoassay test requires that adequate sample volume is available to solubilize the conjugate and to facilitate capillary flow. Insufficient sample volume will lead to limited liquid flow and no "immuno-action". A failed test will result. An example can be found in the cassette type of lateral flow immunoassay device in which a minimum number of drops of specimen is required for test completion. Adding insufficient number of sample drops will lead to erroneous result. Many test manufacturers seek to prevent this problem by painting a pictorial number of sample drops on the cassette to remind the operator the amount of liquid required.

Another example is in saliva drug screen in which the test subject is required to provide specimen sample by swiping a collection swap around in his/her mouth. Due to time constraint and failure of the test subject to follow instruction, insufficient sample often occurs. To address this issue, some manufacturers have added a colored line on the membrane. When sufficient sample is obtained, the colored line will then disappear indicating sufficient liquid has been collected.

10.5.2.3 Reading Under Poor Lighting Conditions

Visual readout is usually required for most lateral flow immunoassay devices. Determination of positive and negative results is based on the operator's perception of the color intensity of the test lines. Hence, this color perception is greatly affected by the ambient light condition. If the test is read in poor lighting conditions, a near positive result (i.e., the specimen contains drug at concentration just below the cut-off) may be interpreted as positive and a near negative result may be interpreted as negative. Hence, false positive and false negative results may occur.

To remove this ambient light issue, it is imperative that operators are trained to read the lateral flow immunoassay test device in a well-lighted area. In addition, several developers have introduced digital readers for lateral flow immunoassay device, which will eliminate this issue altogether (see Chapter 7).

10.5.2.4 Subjective Interpretation

Another consequence of visual read is that the result totally relies on the ability of the operator to perceive the presence of a colored line. The operator must not be color-blind and must be able to perceive the varying intensities of red or blue lines in order to correctly interpret a test result.

For lateral flow immunoassay test manufacturers, the individual variation in color perception must be taken into consideration during the development of the test and its subsequent manufacturing. A test utilizing antibody and antigen pair that gives a sharp dose–response curve helps. The development of a small equipment that is designed to read the test strip is also a solution to this issue (as discussed in the previous section).

10.5.2.5 Reading Too Early or Too Late

Customer requirement dictates that any diagnostic test results should be available as soon as possible and be stable for long period of time. However, the antibody and antigen reaction process in a competitive lateral flow immunoassay is dynamic and the binding of antibody–gold conjugate and the immobilized antigen is a continuous on-and-off process. After a test line has been formed in the membrane, continuous exposure of the membrane with specimen sample (as occurring during the course of the test) may sometimes destabilize the antigen's hydrophobic binding to the membrane leading to the antibody–antigen complex being washed out to the absorbent pad. This may eventually lead to a decrease in color intensity of the test line. Hence, over the course of the test line development, after the application of the specimen, the test line will slowly develop, then the color intensity of the line may continue to get darker or may become lighter depending on the test. This is the reason why manufacturer for each test has chosen an optimal time window to read the test result. This information is available in the product insert of each test kit.

Operators who do not follow the instruction and read the test result either too early or too late will note erroneous results. It is very important for the manufacturer to train the customer to observe the detection time window in order to get the correct result.

10.5.2.6 Using Expired Test Kits

Once manufactured, even under the protective environment enclosed by moisture-barring metallized pouch, lateral flow immunoassay devices will slowly start to deteriorate. Some of the possible contributing factors to the deterioration are:

a. The hydrophobic interactions of the drug–protein conjugate with the nitrocellulose membrane may deteriorate, causing the dislodging of the drug–protein conjugate with the passing of a liquid front. The end result is that a lighter line may be formed.
b. The antibody may degrade over time leading to reduced binding with the drug–protein on the test line and false positive result.
c. The colloidal gold conjugate may be affected by age due to the collapse of the gold colloid around the antibody, causing less intense red color and hence more false positive result.
d. The adhesive on the vinyl backing may lose its binding property over time causing erroneous result (see discussion above and Chapter 3).
e. The pouch seal failed over time causing moisture to enter the pouch and enhance the degradation process.

All these effects suggest that a lighter test and control lines would result. However, in other devices, a darker than normal line may appear after the kit is expired. In any case, false test result would occur if an expired kit is used. To avoid this problem, operator should be trained never to run a test using the device from a kit that has passed its expiration date.

10.5.2.7 Wrong Interpretation and Procedure

For most sandwich-type lateral flow immunoassays (e.g., pregnancy test) or displacement type immunoassay test (e.g., Biosite's Triage® drug screen, San Diego, CA), formation of a colored line indicates a positive result. However, in competitive type assays like most drug screens, the appearance of a test line indicates a negative result. For an operator who has to run both a sandwich test and a competitive assay, or an operator who has just switched from running a

Triage® test to another onsite drug screen, confusion often occur leading to false positive or false negative results.

In addition to the possibility of wrong interpretation, an operator may perform the wrong test procedure. In running a Triage® assay, the operator is required to add the specimen to a reaction well and then add the reaction mixture onto the membrane. However, in a competitive lateral flow immunoassay test, if the urine specimen is applied directly onto the membrane area, disastrous result will occur. The test strip will be flooded and no test or control lines will be formed.

Due to its simplicity, operators often believe that the lateral flow immunoassay is so easy to perform that reading the package insert is a waste of time. Moreover, they may also believe one lateral flow immunoassay is the same as the other. Needless to say, both of these assumptions are incorrect. It is very important that operators must be properly trained prior to their running the lateral flow immunoassay test kit.

10.5.3 Effects of Food, Supplements, and Beverages

Another factor causing false result has been associated with the test subject's consumption of food, food supplements, and beverages that contain the targeted drug or related molecules. In such instances, the lateral flow immunoassay device would identify the specimen from this subject positive. Although the device correctly provides true positive result, such report is often classified as clinically false positive [5] since the test subject was exposed to the targeted drug unintentionally.

One example is the poppy seeds widely used in bakery goods like muffins, bagels, and pie causing positive result in opiate tests. In a report by Rohrig and Moore, after consuming three poppy seed bagels each, urine specimens of four adult volunteers had been shown to contain 314–603 ng per ml of morphine within 4 hours [6]. However, Thevis, Opfermann, and Schanzer had detected much higher concentrations of morphine (more than 1 μg per ml) in the urine of volunteers who had consumed a cake laced with poppy seeds [7].

"Health Inca tea", a popular drink in South America, contains coca leaves. Residual cocaine exists in such beverage and will cause a positive cocaine test [8]. Analysis of a cup of coca tea from two different sources was found to contain over 4 mg of cocaine [9]. In another study, all five subjects who had consumed coca tea were tested positive 2 hours after the consumption, and three of these subjects remained positive after 36 hours. The mean urine concentration was 1,777 ng per ml, which is much higher than the 300 ng per ml cutoff concentration for cocaine [10].

One herbal supplement was also speculated to cause a false negative result for cocaine using a lateral flow immunoassay device. Since this effect was correctable by diluting the urine sample with water, the authors hypothesized that the herbal supplement would complex to the cocaine metabolite (benzylecgonine) and prevented it from being recognized by the antibody. Dilution with water dissociated the complex and allowed the correct test result [11].

No matter how the target drug or its metabolite enters the subject's body and if it is available in his/her body fluid, a positive report by a lateral flow immunoassay device is analytically correct. Test operator needs to be informed of the interference of these specific food and beverage so that appropriate action can be taken in spite of the positive result.

10.5.4 Environmental Factors

The environment may also be a cause of false results. The most common example is the effect of second-hand cigarette smoke. It has been shown that much higher levels of salivary cotinine were detected from nonsmokers from smoking home and workplace than from smoking home and nonsmoking workplace, or from nonsmoking home and smoking workplace. Moreover, nonsmokers from all three situations have cotinine levels at least five times higher than if the nonsmokers are from nonsmoking home and workplace environment [12]. The results suggest the role of environment in interfering with a test result.

However, in the case of marijuana, several studies have shown that in a normal situation, passive inhalation will not lead to a positive marijuana result in both urine and saliva drug screens [5].

Some workplace environment may also contribute to false result. In the preparation of drug detection dog-training aids, laboratory workers under protective garments are routinely exposed to large quantities of methamphetamines. A study by Stout et al. shows that detectable amounts of methamphetamine were found in these workers' urine, suggesting the clinical false positive result due to environmental exposure [13].

10.5.5 Sample Manipulation

False negative results may also occur due to the deliberate manipulation by the test subject. Often, such attempt is found in drug testing. Due to the important impact of a positive test result on one's life, there is a high incentive to defeat a drug test. A comprehensive discussion on this topic has been presented in another volume of this series [14]. Three methods are commonly used to change a positive result to negative. They include substitution, dilution, and chemical adulteration. Commercial products utilizing these methods are available from the internet, magazine advertisement, and "smoke shops".

10.5.5.1 Substitution

This is the process in which the test subject substitutes his/her "dirty" urine with "clean" urine. The "clean" urine can come from another person or a pet. The "dirty" urine can also be purified by an in situ device and resubmitted for testing. Several commercial substitution devices are available and they are quite expensive. Lateral flow immunoassay testing on the substituted specimens will show negative even if the subject is a drug user. Analytically, these specimens are true negative.

10.5.5.2 Dilution

This process involves the consumption of excessive amount of liquids with the purpose of lowering the drug concentration in the urine. The underlying principle for this technique is that since the test result is based on a cut-off concentration, consumption of large amount of liquid will speed the excretion of the target drug from the subject's body and thus lowering its urine concentration hopefully to below the cut-off. Most commercial dilution products consist of large bottle of sweetened water laced with diuretics.

10.5.5.3 Chemical Adulteration

In this technique, exogenous chemicals are added to the urine samples to prevent proper identification of the drugs. Among them, oxidants have been employed to destroy or modify the chemical structures of target drugs. Cross-linking chemicals have also been used to cross-link the drug molecules. The goal of both these methods is to render the drug molecules unrecognizable by the antibodies. Analytically, testing of these adulterated samples by the lateral flow immunoassay will provide a false negative result. Chemical adulterants react very fast (within minutes) and their effects on the target drug molecules are permanent.

Acid, base, or high salt have also been used to change the pH or viscosity of the specimen sample so that the optimal biological environment for the antibody/antigen reaction is no longer available, thus affecting the test result. Unlike the oxidants and cross-linkers targeting the drug molecule, and thus only the test lines in the lateral flow immunoassay are affected, acid, base, and high salt tend to affect both the control and the test lines.

Onsite dipstick test devices are available to determine parameters like pH, specific gravity, creatinine concentration, and the presence of chemicals like oxidants and cross-linkers. Use of such test devices will help to resolve the false negative results [15].

10.5.6 Cross-Reactivity

Cross-reactivity is easily the most discussed topic in lateral flow immunoassay, especially in drug testing. Due to the significant consequence of a positive drug test result, many test subjects that were tested positive in a drug screen often try to find reasons to justify the positive result. Abundant numbers of websites exist that provide information on this issue [16]. Usually, these sites, along with selling sample adulteration products, list drugs or food that would cause false positive result. Unfortunately, some of these information were based on old data and do not reflect the improvement made to the test devices especially since most current immunoassays are being manufactured with monoclonal antibodies instead of polyclonal antibodies as in 1980s and 1990s. Another problem of these websites is that some of the information are based on hearsay and do not supported by scientific data. Moreover, many of the reported "false positives" were produced by various testing technologies that were undefined on the list. Currently, quite a few technologies are being used in drug testing, and they include lateral flow immunoassays, homogeneous immunoassays like CEDIA® (Microgenics, Inc., Fremont, CA) and EMIT® (Dade Behring, Inc., Cupertino, CA), heterogeneous immunoassays like ELISA screens (Immunalysis Corporation, Pomona, CA), and chromatographic and spectroscopic systems like GC/MS or LC/MS. Substances that cause interference on one technology may not present a problem with another technology. Even among lateral flow immunoassays, depending on the antibody and antigen used, interference encountered by one drug screen product may not show up in another manufacturer's product. Often, many reported cross-reacting drugs do not post any problem at all. For example, in an interference study of three drugs on four commercial drug screen kits including Rapid Drug Screen® (American BioMedica Corporation, Kinderhook, NY), DTX® One-Step Drug Test (1 Step Detect Associates, Pittsburgh, PA), DrugCheck® Cup (Express Diagnostics, Inc., Blue Earth, MN), and ToxCup® (Branan Medical Corporation, Irvine, CA), none of the drug screen devices showed false marijuana (THC) positive result for pseudoephedrine at 50 μg per ml in urine. With naproxen at a concentration of 75 μg per ml, all devices displayed lighter THC negative lines but were clearly negative. When somac (pantoprazole) at 75 μg per ml of urine was added to these four devices, only one device showed a definite false positive result while the other three were not affected (unpublished data).

In this section, we shall examine the different causes of this drug-induced false test result. A list of the drugs having shown to affect the lateral flow immunoassay test results and supported by published articles is presented in Table 10.4. Some potentially problematic drugs are also included in the list.

10.5.6.1 Prescription Drug–Induced Positive Results

Many legitimate prescription drugs can cause clinical false positive results. In some instances, the prescription drug is a component of the target drug.

Table 10.4 Table on cross-reactivity

Drugs	Affected drug	Component	Metabolite	X-react
Adderall®	AMP	X		
Ambenyl with Codeine®	OPI	X		
Amphetaminil	AMP		X	
Astramorph PF®	OPI	X		
Avinza®	OPI	X		
Benzphetamine (Didrex®)	MET/AMP		X	
Clobenzorex	AMP		X	
Codimal PH7 Syrup®	OPI	X		
Depodur®	OPI	X		
Desoxyn®	MET	X		
Dexedrine®	AMP	X		
DextroStat®	AMP	X		
Dimethylamphetamine	MET/AMP		X	
Donnagel-PG®	OPI	X		
Duramorph®	OPI	X		
Ethylamphetamine	AMP		X	
Famprofazone	MET/AMP		X	
Fencamine	MET/AMP		X	
Fenethylline	AMP		X	
Fenproporex	AMP			X
Fioricet and Codeine®	OPI		X	
Fiorinal with Codeine®	OPI	X		
Furfenorex	MET/AMP			X
Guiatuss A.C.®	OPI	X		
Infantol Pink®	OPI		X	
Kadian®	OPI	X		
Kaodene with Codeine®	OPI	X		
Kaodene with Paregoric®	OPI	X		
Marinol®	THC	X		
Mefenorex	AMP	X		
MS Contin Tablets®	OPI			X
Oramporph SR®	OPI	X		
Pareforic®	OPI	X		
Phenaphen with Codeine®	OPI	X		
Quiagel PG®	OPI	X		
Robitussin-DAC®	OPI	X		
Roxanol®	OPI	X		
Selegiline (Eldepryl®)	MET/AMP			X
Triacin-C®	OPI	X		
Tylenol with Codeine	OPI	X		
Vicks Inhaler®	MET		X	

In other instances, the prescription drug is metabolized into the target drug. The article from Kong [5] and the SAMHSA's Manual for Medical Review Officer [17] provide an extensive list of prescription drugs that may cause positive results.

a. The prescription drug is a component of the targeted drug being tested: Clinical false positive drug screen results can be caused by prescription drug formulations containing the target drugs as components. (e.g., codeine cough medicine causes the opiate test positive; Adderall® and Marinol® contains amphetamine and Δ9 THC-COOH, respectively) [5].
b. The prescription drug is metabolized into the targeted drug: Sometimes, a prescription drug is metabolized into the target drug (e.g., Selegiline to *l*-methamphetamine and *l*-amphetamine, and Clobenzorex to *d*-amphetamine) [5]. Prenylamine, a calcium antagonist, has also been shown to metabolize into amphetamine, and urine amphetamine concentrations up to 1,280 ng per ml had been detected after a single dose of prenylamine [18].

10.5.6.2 False Positives Due to Cross-Reactivity with a Prescription Drug

Some prescription drugs that are neither components of the target drugs or metabolized into the target drugs may cause cross-reactivity problem. The two abused drugs amphetamine and methamphetamine are especially susceptible to cross-reaction due to their simple single-ring chemical structures. Some of the drugs that have been determined to cause false results are briefly described below.

Venlafaxine (Effexor) is an antidepressant that bears little structural similarity to phencyclidine (PCP), aside from possessing phenyl and cyclohexyl groups. However, both venlafaxine and its metabolite, *o*-desmethylvenlafaxine, has been shown to cause false positive results for two phencyclidine drug screens [19, 20].

Gemfibrozil (Lopid) used in the treatment of hypertriglyceridemia was shown to cause a false marijuana positive result on one lateral flow immunoassay device but not on another. It was speculated that the hydroxymethyl and carboxyl metabolites were responsible for the cross-reaction [21].

Both ephedrine (a bronchodilation medicine) and pseudoephedrine (a nasal decongestant) are structurally related and easily converted to methamphetamine. At high dosages, these medicines may give false positive results for some, but not all, methamphetamine urine drug screens (unpublished data).

In a study on eight commercial onsite screening devices, Leino et al. found that the opioid cough suppressant Pholcodine gave false positive result with amphetamine [22].

Methylphenidate (Concerta or Ritalin), a prescription drug to treat Attention Deficit Hyperacitivity Disorder, has been reported to cause a false positive urine amphetamine test due to cross-reactivity [23].

Efavirenz (Sustiva®), an anti-HIV drug, has been shown to produce false marijuana test result with one drug screen [24]. However, of three other lateral flow immunoassays tested, none gave a false result (unpublished data). Atripla, a HIV drug cocktail, contains Efavirenz and thus may potentially cause false THC result.

Rifampin, a bactericidal antibiotic, was cited by Raher et al. to interfere with two lateral flow opiate immunoassays [25].

10.5.6.3 False Results Reported on Other Immunoassays

Drugs that are associated with false results on immunoassays other than lateral flow immunoassays may post a potential cross-reactivity problem. Some of the reports on these drugs are briefly described below. We have not included reports that appear to be specific to one non-lateral flow immunoassay or reports that were published prior to 2000.

There had been some reports about false positive urine test results for THC in the patients administered with proton pump inhibitors including pantoprazole (Protonix). However, lateral flow immunoassay was not the device being tested [26].

Although no report has been published on the cross-reactivity of quetiapine (a drug to treat schizophrenia) with methadone using onsite drug screen, citations are available that this drug caused false positive results on the tests performed with homogeneous immunoassay systems [27]. Quetiapine has also been associated with false tricyclic antidepressant positive results in a lateral flow immunoassay device and a homogeneous immunoassay test [28].

Fluoroquinolone, a family of broad-spectrum antibiotics, has been cited in several reports to cause false positive opiate drug test results [29].

10.6 Conclusion

False results in lateral flow immunoassay tests do occur. However, they can be reduced by using a good antibody that has been thoroughly evaluated for cross-reactivity and paired with an antigen that can offer a sharp dose–response curve, especially at the dose region around the cut-off concentration. A good operator-training program stressing not only on the test procedures but also on sample manipulation issue would help. Moreover, operators should be informed of the various drugs, food, beverages, and environmental factors that may affect the test results. Only by addressing these issues and with the help of a test reader instrument, lateral flow immunoassay can be utilized in a more quantitative manner.

References

1. Mandatory guidelines and proposed revisions to mandatory guidelines for federal workplace drug testing programs. April 13, 2004 (69 FR 19644).
2. Peace, M., Tarnai, L. and Poklis, A. (2000) Performance evaluation of four On-site drug-testing devices for detection of drugs of abuse in urine. J. Anal. Toxicol. 24:589–594.

3. Moody, D.E., Fang, W.B., Andrenyak, D.M. and Monti, C. (2006) A comparative evaluation of the Instant-View 5-panel test card with OnTrak TesTcup Pro 5: comparison with gas chromatography-mass spectrometry. J. Anal. Toxicol. 30:50–56.
4. Melanson, S.E.F., Lewandrowski, E.L., Griggs, D.A. and Flood, J.G. (2007) Interpreting tricyclic antidepressant measurements in urine in an emergency department setting: comparison of two qualitative point-of-care urine tricyclic antidepressant drug immunoassays with quantitative serum chromatographic analysis. J. Anal. Toxicol. 31:270–275.
5. Kong, T.C. (2008) Clinical false-positive drug test results. In: "Handbook of drug monitoring methods". Dasgupta, A., ed. Humana Press, Totowa, NJ, pp. 395–406.
6. Rohrig, T.P. and Moore, C. (2003) The determination of morphine in urine and oral fluid following ingestion of poppy seeds. J. Anal. Toxicol. 27:449–452.
7. Thevis, M., Opfermann, G. and Schanzer, W. (2003) Case report: urinary concentrations of morphine and codeine after consumption of poppy seeds. J. Anal. Toxicol. 27:53–56.
8. Practical challenges to positive drug tests for marijuana, Editorial. (2003) Clin. Chem. 49:1037–1038.
9. Jenkins, A.J., Llosa, T., Montoya, I. and Cone, E.J. (1996) Identification and quantification of alkaloids in coca tea. Forensic Sci. Int. 77:179–189.
10. Mazor, S.S., Mycyk, M.B., Wills, B.K., Brace, L.D., Gussow, L. and Erickson, T. (2006) Coca tea consumption causes positive urine cocaine assay. Eur. J. Emerg. Med. 13:340–341.
11. Bowen, R., George, D. and Hortin, G. (2005) False-negative results for cocaine metabolites on a lateral-flow drug test slide corrected by dilution. Clin. Chem. 51: 790–791.
12. Tomkins, B.A., Van Berkel, G.J., Jenkins, R.A. and Counts, R.W. (2006) Quantitation of cotinine in nonsmoker saliva using chip-based nanoelectrospray tandem mass spectrometry. J. Anal. Toxicol. 30:178–186.
13. Stout, P.R., Hron, C.K., Klette, K.L. and Given, J. (2006) Occupational exposure to methamphetamine in workers preparing training aids for drug detection dogs. J. Anal. Toxicol. 30:551–553.
14. Wong, R.C. and Tse, H.Y. (2005) Adulteration detection by Intect® 7. In: "Drugs of abuse: Body fluid testing". Wong, R.C. and Tse, H.Y., eds. Humana Press, Totowa, NJ, pp. 233–245.
15. Peace, M.R. and Tarnai, L.D. (2002) Performance evaluation of three on-site adulterant detection devices for urine specimens. J. Anal. Toxicol. 26:464–470.
16. http://www.passyourdrugtest.com/false-positives.htm
17. Medical Review Officer Manual for Federal Agency Workplace Drug Testing Programs. US Department of Health and Human Services, Substance Abuse and Mental Health Services Administration, Division of Workplace Programs. Available at: http://dwp.samhsa.gov/DrugTesting/Level_1_Pages/HHS%20MRO%20Manual%20(Effective%201, %202004. Accessed 5/12/2008.
18. Kraemer, T., Roditis, S., Peters, F. and Maurer, H. (2003) Amphetamine concentrations in human urine following single-dose administration of the calcium antagonist Prenylamine—Studies using fluorescence polarization immunoassay (FPIA) and GC-MS. J. Anal. Toxicol. 27:68–73.
19. Sena, S., Kazimi, S. and Wu A. (2002) False-positive phencyclidine immunoassay results caused by Venlafaxine and O-Desmethylvenlafaxine. Clin. Chem. 48:676–677.
20. Santos, P.M., Lopez-Garcia, P., Navarro, J.S., Fernandez, A.S., Sadaba, B. and Vidal, J.P. (2007) False positive phencyclidine results caused by Venlafaxine. Am. J. Psychiatry 164:349.
21. Lewis, J.H. (1999) Interference of Gemfibrozil with Roche Testcup. J. Anal. Toxicol. 23:384.
22. Leino, A., Saarimies, J., Gronholm, M. and Lillsunde, P. (2001) Comparison of eight commercial onsite screening devices for drugs-of-abuse testing. Scand. J. Clin. Lab. Invest. 61:325–331.

23. Manzi, S., Law, T. and Shannon, M.W. (2002) Methylphenidate produces a false-positive urine amphetamine screen. Pediat. Emerg. Care. 18:401.
24. Rossi, S., Yaksh, T., Bentley, H., van den Brande, G., Grant, I. and Ellis, R. (2006) Characterization of interference with 6 commercial Δ^9-tetrahydrocannabinol immunoassays by Efavirenz (Glucuronide) in urine. Clin. Chem. 52:896–897.
25. Daher, R., Haidar, J.H. and Al-Amin, H. (2002) Rifampin interference with opiate immunoassays. Clin. Chem. 48:203–204.
26. http://www.fda.gov/cder/foi/label/2004/20987slr020_protonix_lbl.pdf. Accessed 5/18/2008.
27. Widschwendter, C.G., Zernig, G. and Hofer, A. (2007) Quetiapine cross reactivity with urine methadone immunoassays. Am. J. Psychiatry 164:172.
28. Hendrickson, R.G. and Morocco, A.P. (2003) Quetiapine cross-reactivity among three tricyclic antidepressant immunoassays. J. Anal. Toxicol. 41:105–108.
29. Zacher, J. (2004) False-positive urine opiate screening associated with Fluoroquinolone use. Ann. Pharmacothe. 38:1525–1528.

Chapter 11
Regulatory Issues in the Development and Marketing of Lateral Flow Immunoassays

Raphael C. Wong

11.1 Introduction

As with all other medical devices, commercializing a lateral flow immunoassay device requires special attention to regulatory issues. Negligence may prevent the device from being marketed no matter how good the product is. This chapter will address some of these regulatory issues as encountered in the United States and the rest of the world. A flow chart of the pertinent processes is presented in Fig. 11.1. Since regulations are constantly being updated, the issues discussed only reflect the common practices at the time of publication of this book. It is prudent that a device developer should have adequate resources in both regulatory and legal arenas to address these issues and to do it vigilantly.

11.2 Patent Considerations

Prior to the commencement of any research and development effort on a lateral flow immunoassay, literature search on the assay should include patent search of all relevant prior art. This should be done in addition to the normal academic and industrial literature searches. A cursory patent search can begin by examining the related patents on the database of the US Patent and Trademark Office (PTO) via the website www.uspto.gov/ and the European Patent Office database at http://www.esp@cenet.com

According to the US Patent and Trademark Office, there are three types of patents: utility, design, and plant. Medical devices, including the lateral flow immunoassay, involve the utility and design patents and not the plant patents. Utility patents are new and useful proprietary matters related to process, machine of article of manufacture, or composition of matter. They also include

R.C. Wong (✉)
Branan Medical Corporation, 140, Technology Drive, Suite 400, Irvine, CA, USA
e-mail: raphael@brananmedical.com

R.C. Wong, H.Y. Tse (eds.), *Lateral Flow Immunoassay*,
DOI 10.1007/978-1-59745-240-3_11, © Humana Press, New York, NY 2009

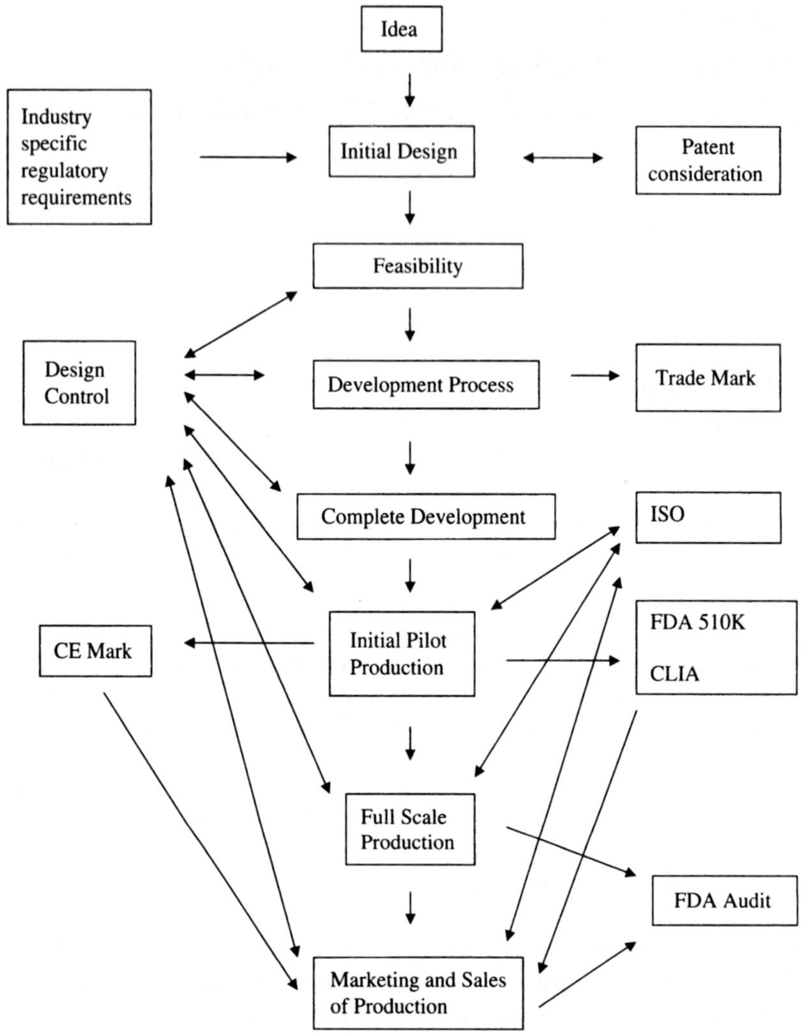

Fig. 11.1 Summary of regulatory issues in the United States

any new and useful pertinent improvements. Design patents, on the other hand, are related to new, original design for an article that can be manufactured.

Under the Patent Act, an invention must be new, useful, and non-obvious to be patentable. However, invention with the following characteristics is not patentable [1]:

(1) Prior art is available. This includes invention that has been known to the public; or it has been described in a publication; or it has been used publicly.
(2) The applicant for the invention has offered the innovative product for sale to the public or has published the invention for more than one year prior to the filing date. This one-year grace period only applies to the United States In other countries, usually the one-year grace period is not allowed.

In the field of lateral flow immunoassay, many patents do exist that cover the technologies, formats, and raw materials being used in the engineering and manufacturing of the assay. It is beyond the scope of this chapter to detail the patents and the law suits involved, readers can obtain a copy of Millipore, Inc.'s publication on "Rapid Lateral Flow Test Strips" [2], which listed some pertinent patents on lateral flow immunoassay to understand certain aspects of the current patent situation. Since litigation is very active in this field, the patent literature should be thoroughly reviewed and appropriate legal considerations taken to prevent possible legal problem after product launch. In certain circumstances, it may be advantageous to license appropriate patented technologies from license holders. However, this would definitely increase the cost of the final product as most licenses will require an upfront fee and a running royalty payment.

In the event that the technology, format, or raw materials being developed are deemed to be novel after in-depth reviews of the patent literature, a patent application may be filed with the patent office. Usually, this is done with the help of a patent attorney. It is important to choose the patent attorney carefully since prosecuting a patent application is expensive. Using an attorney who is experienced in the specific field of the lateral flow immunoassay technology can cut the cost of the preparation of the patent application by eliminating the cost associated with educating him/her on the background information. With an attorney knowledgeable in the field, a good patent text and all appropriate claims will be included in the application. More importantly, when the patent office examiner has specific questions or raises objections to certain claims, an experienced patent attorney would be better equipped to deal with such situations.

Once the attorney has been chosen and the application prepared, the countries where the patent protection are needed have to be determined. US inventors should recognize that the current stage of global economy offers market opportunity for lateral flow immunoassays worldwide and patent protection should be sought in territories in addition to the United States. However, since each country may require a specific application format (requiring additional legal fee), separate filing fee, prosecuting expense, and patent maintenance fee, it is prudent that applications should be filed to countries where product sales are anticipated.

At this stage, depending on the availability of funds and the significance of the invention, either one of the two types of patent applications is available. A standard patent application is one that contains a detailed written description of the invention, drawings, and claims. Filing a standard patent application does not guarantee the grant of a patent. The granting is dependent upon the outcome of an examination by a reviewer from the patent office. Also, there is generally a long delay between the filing date and the examination date.

Sometimes, a provisional patent application that is inexpensive and less complex may be filed. It provides the opportunity for the inventor to place an application on file with the patent office to obtain a filing date (priority date). Such priority date becomes important in a patent litigation of two companies holding similar patents in deciding who has the idea first. A provisional patent allows the inventor to designate the invention "patent pending" for a full year in the United States. This means the inventor is able to discuss his/her invention with others, evaluate its commercial potential, and seek funding, all while being protected. The disclosure in a provisional application can be incorporated into a standard patent application if a patent is to be pursued. If a follow-up standard patent application is not filed, the provisional application expires. There is no enforceable rights through the filing of a provisional application. Prior to June 8, 1995, patents were normally issued for a non-renewable period of 17 years, measured from the date of issuance. Under the current provision, the term is now 20 years measured from the date of application.

11.3 Industry-Specific Regulatory Requirements

When developing lateral flow immunoassay for a particular field, additional regulatory requirements must be considered, especially in the specialized onsite drug screen industry. For example, the main US government agency involved in the drugs-of-abuse testing is the Substance Abuse and Mental Health Service Administration (SAMHSA) of the Department of Health and Human Service. SAMHSA provides guidelines on drug testing for federal employees [3]. Among the information included on the guidelines are: what specimen matrices are allowed; what procedure is involved in the collection, testing, and storage of the specimens; which drugs are to be tested; and what are their cut-off concentrations. Developer of drug screens for the US workplace drug testing market should take into consideration the SAMHSA guidelines and ensure that the drug screen products conform to them, since these guidelines are being adopted by most users of drug test products in the testing of their employees.

Australia and New Zealand also have jointly published "Procedures for specimen collection and the detection and quantitation of drugs of abuse in urine" [4]. Similar to the US guideline, this joint standard, prepared by the Standards Australia/Standards New Zealand Committee CH-036, stipulates the drugs and cut-off levels for drug screen. However, the drugs chosen and the levels are somewhat different from the SAMHSA guidelines.

Other countries like the United Kingdom and Japan have also developed their own drug testing guidelines that are again somewhat different from those of the United States. All these information should be taken into consideration prior to embarking on developing drug screens catering to these markets.

11.4 Design Controls

For US assay developers, it is important that the product development process follows the design control guidelines stipulated by the Food and Drug Administration (FDA) [5]. Failure to do so may lead to the issuance of a warning letter during an FDA inspection.

Design controls describe an interrelated set of procedures serving as a system of checks and balance incorporated into the assay development process. This systematic assessment of the design helps to discover in the early development stages the deficiencies in design and discrepancies between the proposed designs and the product requirements. Such a system involves the participation of managers of various functional departments in the manufacturing organization. Advantages of the design control system include:

a) Enabling the final product to be appropriate for its intended use;
b) Providing a forum for managers and designers to communicate with each other;
c) Helping to discover any problems in an early stage, thus allowing corrections to be made and adjusting the allocation of resources.

In practice, the system enables the assay development process to proceed in a sequence of phases in which requirements are developed and a device is designed to meet those requirements. The design is then evaluated and transferred to production, and the device is manufactured. Feedback paths are established between each phase of the process and previous phases. Specifically, components of the system include:

a) Design and development planning
b) Design input
c) Design output
d) Design review
e) Design verification
f) Design validation
g) Design transfer
h) Design changes
i) Design history file.

Due to the scope of the design control process and the necessity of the involvement of every functional departmental manager, design control is expensive and time-consuming. However, it is mandatory for US manufacturers. Most importantly, the design control process helps to tackle the development issues in detail far in advance so as to eliminate the potential problems when the product is launched to the marketplace.

11.5 Trademark

Applying for regulatory clearance to market a specific lateral flow immunoassay device requires the device to be identified by a unique name called trademark. Trademark is an important business asset because it allows the manufacturer to establish the lateral flow product reputation without having to worry that an inferior “knock off” product with the same name will come onto the market later, resulting in diminished reputation or lower profit. Hence, during the later development phase of the device, trademark application for the device name should be established.

In the United States, trademark rights are established by either actually using the mark in marketing the lateral flow assay product or filing a proper application to register a mark with the Patent and Trademark Office (PTO) [6]. Although registering with the PTO is not mandatory for establishing trademark rights, such a process is beneficial because of the following:

a) It provides an official notice of the claim to the mark.
b) It shows evidence of ownership.
c) It allows the capability of invoking federal court jurisdiction.
d) It establishes a basis for obtaining registration in foreign countries.
e) It prevents the importation of infringing foreign goods.

Similar to the patent application, the first step in the application process involves a search (trademark search) to ascertain that the proposed trademark is unique. One can search for registered trademarks on the PTO's website: www.uspto.gov/main/sitesearch.htm. Alternatively, a professional trademark search company can be retained to perform the search. The benefit of utilizing such a company is that it can research not only federally registered marks but also common law trademarks.

Once the trademark search shows that the proposed mark is unique, information such as product category, description of use, and labels where the trademark is incorporated are filed along with the application to the PTO. With the trademark being filed, the specific mark can be identified with the "TM" symbol. The registration process generally takes 6 months from start to finish. The PTO will usually assign a serial number for the filing about 2 months. After passing the scrutiny of the PTO, the mark is then published in the Official Gazette. Other parties will then have 30 days from the publication date to object to the mark. If no objection is encountered, the PTO will either issue a Certificate of Registration (if the application is based on actual use of the assay device on the market) or a Notice of Allowance (if the application is based on intended use). A registered trademark would be identified with the "®" symbol. For obvious reason, it is important to apply for a device trademark as early as possible so that a unique and appropriate name can be secured to prevent other manufacturers from using the same name.

Manufacturers that market their products in foreign countries should consider applying trademarks in those countries to prevent the possible scenario that these products cannot be marketed under their established US names. It is currently not possible to file and obtain a single trademark registration that will automatically apply around the world. Like any national law, trademark laws apply only in their applicable country or jurisdiction.

11.6 Product Clearance in the United States

For a lateral flow immunoassay product to be marketable in the United States, clearance by the US FDA is required [7]. According to the US Federal Food, Drug, and Cosmetic Act, lateral flow immunoassay product is a medical device that falls into the In-Vitro Diagnostic (IVD) category, defined as the chemicals and reagents used for testing biological fluids for the aid in diagnosis of human diseases.

There are three classes of IVD devices. Class I devices are low-risk devices. They are generally exempt from premarket review by the FDA prior to being marketed to the public. However, these products still are subject to general controls such as Good Manufacturing Practice regulations, facility registration, and device listing.

Class II devices are of intermediate risk and generally require Pre-market Notification submission known as the "510(k)". Most lateral flow immunoassays are classified as Class II devices. Submission of a 510(k) for a device requires documenting the new device is substantially equivalent to an existing, legally marketed device known as the "predicate" device. Much of the information about a predicate device can be found in the package insert of the predicate device. Some additional relevant information of the predicate device may be available from the 510k summary or statement from the FDA website (www.fda.gov/cdrh/510kdatab). More detailed information about the predicate kit can be obtained through "freedom of information" notification to the FDA for the device's 510(k) filing information. For Class II devices, in addition to the general controls, FDA subjected them to special controls such as FDA guidance documents, device tracking, patient registries, and post-marketing clinical studies.

Class III devices are for the devices that have no predicate kits. They are stringently regulated and subject to all general and special controls. Ability to market a product depends on the approval by FDA of a Pre-market Approval Application (PMA) submission. In this case, multi-center trials of appropriate patient population of a statistically justifiable sample size are required. The results obtained must be shown to be intrinsically safe and effective through rigorous statistical analysis. Because of the resources required, it has been estimated that a PMA would increase the cost of filing by approximately 20-fold when compared to a 510(k) filing [8].

It is essential that a manufacturer should establish what classification is applicable for its device in the initial phase of research and perform the required data to support the specific application.

11.7 CLIA Waived and Over the Counter Products

Along with the product clearance, FDA is also responsible for classifying the complexity of the lateral flow immunoassay device so as to determine the appropriate personal training required to perform the device under the Clinical Laboratory Improvement Amendments of 1988 (CLIA) [9]. The three classifications of high complexity, moderate complexity, or waived complexity would determine whether the device can be administered by a trained professional, a less experienced laboratory technician, or by a non-technician [10]. Appropriate quality control procedures are required for all products. The most desirable classification for a product is the CLIA waived category. This includes devices that are so simple and accurate as to render the likelihood of erroneous results negligible; or they pose no reasonable risk of harm to the patient if the tests are performed incorrectly. Thus, CLIA waived products can be used in non-laboratory setting. Some of the lateral flow immunoassays that have been cleared under this category include ovulation tests, urine pregnancy tests, influenza tests, and urine drug screens. It is important to understand that CLIA waiver clearance is brand name–specific such that one company's drug screen may have CLIA waiver while another company's drug test product may not have such designation. Other than the few CLIA waived devices, most lateral flow immunoassays are designated as moderately complex. They are to be used by professionals.

Application for the waived status usually is done after the product has received FDA clearance. To obtain such a status, it is stipulated that manufacturers must conduct a series of studies aiming at demonstrating that the device will provide the same accuracy whether it is administered by a non-skilled person or by a trained professional.

If the devices are to be marketed over-the-counter (OTC) to consumers for home use, then the manufacturer should apply for an OTC license. This is also done after the product clearance. Products approved for home use are automatically qualified for CLIA waiver.

If the lateral flow immunoassay product market is in the United States, it is important to pay attention to CLIA and OTC regulations. With the current trend of shifting diagnostic testing function to satellite laboratories, clinics, and outpatient offices along with the increasing dependence on patient self-testing, achieving CLIA waiver status or OTC clearance for a product is very desirable.

11.8 CE Marking

If a lateral flow immunoassay product market covers the member states of the European Union (EU) and the European Free Trade Area (DFTA), a CE marking must be affixed to the device to indicate conformity with the essential health and safety requirements set out in the European In-vitro Diagnostic Directive (IVDD). With a CE Marking, a device can be legally placed on and transport within the European market. In the event that the device becomes non-performing, the CE marking will help its withdrawal by customs and enforcement authorities. Moreover, the marking indicates to customers that the product meets designated minimum safety standards and it is manufactured with a minimum level of quality. Hence, the CE marking enhances the credibility of the product. Without such a marking, a lateral flow immunoassay device will not be allowed access to the EU market [11].

The first step to obtain CE marking is to assess the classification of the device according to the IVDD. There are four "classes" reflecting the risks involved in the use of the product as follows.

1) Class 1 is for products that offer low risk.
2) Class 2A products are of medium risk and include tests for certain blood groupings and infections with HIV, HTLV, and hepatitis viruses.
3) Class 2B is for tests that are deemed with elevated risk and include tests for blood transfusions, infectious diseases, tumors, and hereditary diseases.
4) Additionally, there is a class of products for self-testing by a non-skilled person in a home environment.

For Class 1 devices, which cover many lateral flow immunoassay devices (e.g., drug screen and pregnancy test), the initial step to obtain a CE marking involves the preparation of a technical file encompassing design documentation, manufacturing test reports, and operation information to show conformity as required by the IVD directive. Such a technical file is usually prepared by the manufacturer. A European "Authorizing Representative" is then appointed. This representative must have a physical office in Europe and qualified to handle regulatory matters. The Authorizing Representative, which would be identified on the labeling of the lateral flow immunoassay device, will register the product with the "Competent Authority" in the country where the Authorized Representative is based and every European country where the product will be sold. Subsequently, the manufacturer prepares a legally binding document called the "Declaration of Conformity," certifying compliance by declaring that all the CE Marking requirements are being met. This is often refer to the manufacturer "self-certify" its product. Then, the manufacturer can affix CE marking to the device label. Only after completing all these steps can the device be shipped to Europe.

For devices that fall into other classes, the route to CE marking is more complicated. The manufacturer is required to implement a quality system (e.g., the ISO 13485:2003 quality system described in another section of this chapter) as well as preparing the technical file. These documents must then be audited by a "Notified Body" – an organization nominated by a EU member government and notified by the European Commission to provide assessment of the manufacturer's conformity to the requirements of the directive. The manufacturer still needs to appoint an Authorizing Representative. Upon the passing of the audit, the Notified Body will issue a CE Certificate and the manufacturer will prepare a Declaration of Conformity and affix the CE marking.

In addition to acting as the European representative of the manufacturer, the Authorized Representative's responsibilities also include:

1) It collects all possible product problem issues (incidents) from end users of the manufacturer's device.
2) It responds to all device-related complaints that suggest an incident or near incident has occurred and notifies the manufacturer.

3) It may obtain relevant expert opinions in Europe for a severe incident.
4) With the agreement of the manufacturer, it may report the incidents or near incidents to the appropriate Competent Authority.
5) It may procure devices that cause the incident and send them to the manufacturer for evaluation.
6) It would obtain competent legal advice for the manufacturer if necessary.
7) It updates the manufacturer with changes in legislation occurring in the EU, which are relevant to the manufacturer's business.

It is important to choose a good, professional European Authorized Representative, since changing the Authorized Representative would incur great expenses and may cause unforeseeable confusion. Some the expenses include: the removal of existing packaging materials; purchasing new labeling materials; and the requirement to inform the EU authorities and customers of the new representative.

11.9 International Organization for Standardization Certification

In the 1970s, many countries established their own quality standards to gain consumer confidence. With the increase in international trade, the development of internationally recognized quality management standards became essential and eventually leads to the establishment of the International Organization for Standardization (ISO), which is the world's largest standards developing organization.

ISO 9000 is a family of standards for quality management systems. For lateral flow immunoassay device manufacturers, the ISO requirement is encompassed in the ISO 13485:2003 standard. It provides procedures for key processes including:

1) The monitoring of manufacturing processes to ensure quality products are being produced.
2) The keeping of proper records.
3) The checking of finished goods for defect.
4) The installation of corrective action once defect is found.
5) The review of individual processes and the quality system for effectiveness.
6) The facilitation of continual improvement.

ISO, being a standard establishing institution, does not certify organizations. Accreditation Bodies (AB) are formed in many countries to authorize Certification Bodies (CB) for the audit of manufacturers applying for ISO certification. Mutual agreements among various AB allow certificates issued by any of the accredited CB be accepted worldwide.

Manufacturer that applies for an ISO certification should first contact a CB in its region. Since the main role of the CB is to help the manufacturer to meet the quality standards, there would usually be a pre-audit performed by the CB audit to briefly assess the quality system to see if major problem exists and recommend remedy to the problem. Only after both the manufacturer and the CB auditor are satisfied the quality system is sound that the manufacturer would be assessed extensively on its facility, organization, products, services, and processes via a formal audit by the CB. When a problem is encountered, a "deficiency " or "non-compliance" is issued to the management of the manufacturer. Correction is required before a certificate can be issued. If no major problem is encountered, or if the CB receives a satisfactory resolution on the listed problem, an ISO 9001 certificate can then be issued. This certificate must be renewed at intervals usually after a surveillance audit.

The ISO standard also requires the manufacturer to perform self-audits (internal audits) by its trained staff at regular intervals. The goal for all these processes is, via a continual process of review and assessment, to verify that the quality system is working as designed, to discover where improvement can be made, and to correct or prevent problems before they occur.

It is important to realize that an ISO certification may serve as an impressive marketing tool, but if a manufacturer has no actual desire to improve and maintain quality, then the ISO quality system will be just an expensive paper system that is meaningless to the company or the products. Many companies have found utilization of the ISO13485:2003 standard do help to improve the quality of their products.

11.10 FDA Facility Inspection

Once a manufacturer has received the relevant 510 K clearance, it can market the lateral flow immunoassay device immediately in the United States. However, the manufacturer should be prepared for an FDA quality system inspection [12] at any time by a FDA investigator who has the authority to enter the manufacturer's facility to audit equipment, materials, products, labeling, and certain records. The initial inspection of a new facility may be comprehensive. It may also be focused on a specific issue. For an established facility, an inspection may have been triggered in response to a reported problem. The inspector usually observes operations, examines equipment, reviews documents, collects product samples, and interviews employees.

At the conclusion of the inspection, the investigator usually holds a meeting with the management of the facility. If the inspector finds the manufacturer's operation meets FDA requirement, there would be no written observation. However, if the inspector discovers conditions that deviate from FDA requirements, a notice of inspectional observations called the "Form 483" would be issued. The inspector would review each observation and ask for comments. This will allow the management the opportunity to present any corrective actions that had been undertaken during the audit.

Depending on the severity of the violations, the manufacturer may receive an FDA Warning Letter summarizing the findings of the inspection. The manufacturer should prepare a formal response letter detailing answer for each item in the letter or Form 483 with a timeline for correcting the issue and deliver it promptly to the FDA. It is important that immediate actions should be undertaken to correct the deficiencies identified in the Form 483 or the warning letter. FDA may conduct a follow-up inspection in 3–6 months to determine if corrections have been made. Unsatisfactory compliance may lead to enforcement actions ranging from recall and seizure of products, fine, and even criminal liability of senior executive of the manufacturing firm.

11.11 Requirements of Other Countries

Many countries have also established their own regulations on medical devices, which include lateral flow immunoassay products. A couple of examples are briefly described below.

In Canada, the Canadian Medical Devices Conformity Assessment System (CMDCAS) mandates that manufacturers selling class II, III, and IV medical devices to Canada must be registered by the quality systems established by the Standards Council of Canada (SCC), which include ISO13485:2003 standards as part of the system.

Similarly, the Japanese Pharmaceutical Affairs Law (PAL) Regulations require registration to ISO13485:2003 as part of the quality requirement.

11.12 Conclusion

To bring a medical device like a lateral flow immunoassay test to the market properly requires the attention to details of many regulatory issues. They are cumbersome and complex. They also demand the allocation of large amount of resources. And many of these regulations have to be met because they are the law. Many of these regulatory requirements, however, do help to improve the product design and product introduction process, and ensure that a viable and safe product can be manufactured. They also may serve as a barrier to entry to the onsite market for less qualified manufacturers.

References

1. “General Information Concerning Patents”. US Patent and Trademark Office. www.uspto.gov/go/pac/doc/general/
2. “Rapid Lateral Flow Test Strips: Considerations for Product Development”. Millipore Inc., 2006.

3. "Mandatory Guidelines and Proposed Revisions to Mandatory Guidelines for Federal Workplace Drug Testing Programs; Notices". Federal Register, Vol. 69, No. 71 pp. 19644–19673.
4. "Procedures for Specimen Collection and the Detection and Quantitation of Drugs of Abuse in Urine", by Australian/New Zealand Standard. AS/NZS 4308:2008, March 19, 2008.
5. "Design Control Guidance For Medical Device Manufacturers". FDA Center for Devices and Radiological Health, March 11, 1997, www.fda.gov/cdrh/comp/designgd.pdf
6. "Basic Facts about Trademarks". US Patents and Trademark Office. www.uspto.gov/web/offices/tac/doc/basic
7. "Getting to Market with a Medical Device" US Food and Drug Administration. www.fda.gov/CDRH/DEVADVICE/3122
8. Smith, K. and Kates, J. (1966) Regulatory hurdles in bringing an in vitro diagnostic device to market. Clin. Chem. 42(9):1556–1557.
9. "CLIA – Clinical Laboratory Improvement Amendments". US Food and Drug Administration. www.fda.gov/CDRH/clia/
10. "Public Health Service; CLIA Program; Categorization of Waived Test". Department of Health and Human Services, Health Care Financing Administration, Federal Register 42 CFR 493.15 (b) and 42 CFR 493.15 (c), September 13, 1995, Vol. 60, No. 177, pp. 47534–47543.
11. "Guidance Notes on In Vitro Diagnostic Medical Devices Directive 98/79/EC". Competent Authority (UK), February 2006, www.mhra.gov.uk/Howweregulate/Devices/InVitroDiagnosticMedical DevicesDirective/index.htm
12. "Food and Drug Administration, Title 21, Part 820 Quality System Regulation". 21CFR820.100

Index

Printed in the United States of America

Lightning Source UK Ltd.
Milton Keynes UK
UKOW031614270513

211324UK00010B/116/P

9 781588 299086